R0021935456

D0844251

RY
ENTER

R0021935456

REF
QD 471 .B316 Cop.1

Basolo, Fred, 1920-.
Coordination chemistry

DATE			

DISCARD

REF
QD
471
.B316
Cop.1

FORM 125 M

Business/Science/Technology Division

The Chicago Public Library

Received FEB 16 1979

© THE BAKER & TAYLOR CO.

Coordination Chemistry

THE GENERAL CHEMISTRY MONOGRAPH SERIES

Russell Johnsen, Editor
Florida State University

Gordon M. Barrow (*Case Institute of Technology*)	THE STRUCTURE OF MOLECULES
Fred Basolo (*Northwestern University*), Ronald C. Johnson (*Emory University*)	COORDINATION CHEMISTRY
Gregory R. Choppin (*Florida State University*)	NUCLEI AND RADIOACTIVITY
Werner Herz (*Florida State University*)	THE SHAPE OF CARBON COMPOUNDS
Robin M. Hochstrasser (*University of Pennsylvania*)	BEHAVIOR OF ELECTRONS IN ATOMS
Edward L. King (*University of Colorado*)	HOW CHEMICAL REACTIONS OCCUR
Bruce H. Mahan (*University of California*)	ELEMENTARY CHEMICAL THERMODYNAMICS

Coordination Chemistry

The Chemistry of Metal Complexes

Fred Basolo
Northwestern University

Ronald C. Johnson
Emory University

W. A. BENJAMIN, INC.,
Menlo Park, California · Reading, Massachusetts
London · Amsterdam · Don Mills, Ontario · Sydney

COORDINATION CHEMISTRY
The Chemistry of Metal Complexes

REF
QD
471
.B316
Cop. 1

Copyright © 1964 by W. A. Benjamin, Inc.
All rights reserved

Library of Congress Catalog Card Number 64-22273
Manufactured in the United States of America

ISBN 0-8053-0651-X
GHIJKLMNOP-AL-7987

*Final manuscript was received on January 30, 1964;
this volume was published on September 14, 1964*

The publisher wishes to acknowledge the assistance of William Prokos, who produced the illustrations

W. A. BENJAMIN, INC., Menlo Park, California

BUSINESS/SCIENCE/TECHNOLOGY DIVISION
THE CHICAGO PUBLIC LIBRARY

FEB 16 1979

Dedication

This book is dedicated to Professor John Christian Bailar, Jr., our professional father and grandfather. Through his research and his training of students Professor Bailar is in large part responsible for the development of coordination chemistry in America.

Editor's Foreword

THE TEACHING OF GENERAL CHEMISTRY to beginning students becomes each day a more challenging and rewarding task as subject matter becomes more diverse and more complex and as the high school preparation of the student improves. These challenges have evoked a number of responses; this series of monographs for general chemistry is one such response. It is an experiment in the teaching of chemistry which recognizes a number of the problems that plague those who select textbooks and teach chemistry. First, it recognizes that no single book can physically encompass all the various aspects of chemistry that all instructors collectively deem important. Second, it recognizes that no single author is capable of writing authoritatively on *all* the topics that are included in everybody's list of what constitutes general chemistry. Finally, it recognizes the instructor's right to choose those topics that he considers to be important without having to apologize for having omitted large parts of an extensive textbook.

This volume, then, is one of approximately fifteen in the General Chemistry Monograph Series, each written by one or more highly qualified persons very familiar with the current status of the subject by virtue of research in it and also conversant with the problems associated with teaching the subject matter to beginning students. Each volume deals broadly with one of the subdivisions of general chemistry and constitutes a complete entity, far more comprehensive in its coverage than is permitted by the limitation of the standard one-volume text. Taken together, these volumes provide a range of topics from which the individual instructor can easily select those that will provide for his class an appropriate coverage of the material he considers most important.

Furthermore, inclusion of a number of topics that have only recently been considered for general chemistry courses, such as thermodynamics, molecular spectroscopy, and biochemistry, is planned, and these volumes will soon be available. In every instance a modern structural point of view has been adopted with the emphasis on general principles and unifying theory.

These volumes will have other uses also: selected monographs can be used to enrich the more conventional course of study by providing readily available, inexpensive supplements to standard texts. They should also prove valuable to students in other areas of the physical and biological sciences needing supplementary information in any field of chemistry pertinent to their own special interests. Thus, students of biology will find the monographs on biochemistry, organic chemistry, and reaction kinetics particularly useful. Beginning students in physics and meteorology will find the monograph on thermodynamics rewarding. Teachers of elementary science will also find these volumes invaluable aids to bringing them up to date in the various branches of chemistry.

Each monograph has several features which make it especially useful as an aid to teaching. These include a large number of solved examples and problems for the student, a glossary of technical terms, and copious illustrations.

The authors of the several monographs deserve much credit for their enthusiasm which made this experiment possible. Professor Rolfe Herber of Rutgers University has been of invaluable assistance in the preparation of this series, having supplied editorial comment and numerous valuable suggestions on each volume. Thanks are also due to Professor M. Kasha of the Florida State University for many suggestions during the planning stages and for reading several of the manuscripts.

RUSSELL JOHNSEN

Tallahassee, Florida
October 1962

Preface

COORDINATION CHEMISTRY is primarily concerned with metal complexes, but many of its concepts are applicable to chemistry in general. Beginning students, therefore, will profit from an appreciation and understanding of the basic principles of coordination chemistry, which may later be applied in more sophisticated fashion in advanced courses.

Although textbooks of general chemistry usually contain brief treatments of metal complexes and coordination chemistry, their limited space precludes the discussion of many of the important aspects of the subject. This being so, the present book was written to supplement the material now available on the subject to freshman chemistry students. The authors believe that the material presented will also be of value to students in junior-senior level courses in inorganic chemistry.

Modern theoretical concepts as applied to metal complexes are used. At first glance, such an approach may seem more difficult and confusing than a somewhat more traditional treatment. However, it is our experience that beginning students are able to grasp these concepts, thus making it easier for them to understand the material as presented in advanced courses. The valence bond theory is mentioned only briefly, whereas the crystal field and molecular orbital theories are discussed in considerable detail. These theories are used to explain the stability and liability of metal complexes. Current theories of reaction mechanisms are also included.

The authors would appreciate both suggestions for the improvement of the book and reports of student reaction toward it. We wish to thank Dr. S. A. Johnson, who read the entire manuscript and made many helpful suggestions. One of us (F. B.) wishes to thank Dr. V. Caglioti and the people in his Institute at the University of Rome, where part of the original writing of this book was done, for their generous help and hospitality.

Fred Basolo
Evanston, Illinois

Ronald C. Johnson
Atlanta, Georgia

July, 1964

Contents

I

Introduction and Historical Development

1-1 INTRODUCTION

At a very early stage in his course work in chemistry the student is introduced to a class of compounds referred to as *coordination compounds*, *metal complexes*, or just *complexes*. These are compounds that contain a central atom or ion, usually a metal, surrounded by a cluster of ions or molecules. The complex tends to retain its identity even in solution, although partial dissociation may occur. It may be a cation or an anion or nonionic, depending on the sum of the charges of the central atom and the surrounding ions and molecules. It is the chemistry of this type of compound that is described in this book.

Coordination compounds play an essential role in the chemical industry and in life itself. The 1963 Nobel prize in chemistry was awarded jointly to Dr. K. Ziegler of the Max Planck Institute, in Germany, and to Prof. G. Natta of the University of Milan, in Italy. Their research was responsible for the development of the low-pressure polymerization of ethylene, which now makes thousands of polyethylene articles commonplace. The Ziegler-Natta catalyst for this polymerization is a complex of the metals aluminum and titanium. The importance of metal complexes becomes clear when one realizes that chlorophyll, which is vital to photosynthesis in plants, is a magnesium complex and that hemoglobin, which carries oxygen to animal cells, is an iron complex.

It is almost certain that you have already encountered coordination compounds in the laboratory or elsewhere. They are used extensively in qualitative analysis as a means of separating certain metal ions and also as a means of positively identifying certain unknown ions. For example, you have perhaps performed the experiment used to identify silver ion in solution. Recall that if silver ion is present, the addition of chloride ion gives an immediate white precipitate of silver chloride. This precipitate dissolves in an excess of aqueous ammonia; but if to the clear solution is added an excess of nitric acid, the white precipitate is formed again. This behavior is due to the equilibria (1), (2).

$$Ag^+ + Cl^- \rightleftharpoons \underset{\text{white}}{AgCl\downarrow} \tag{1}$$

$$AgCl + 2NH_3 \rightleftharpoons \underset{\text{clear solution}}{[Ag(NH_3)_2]^+ + Cl^-} \tag{2}$$

A white precipitate forms as shown in (1) because AgCl is not soluble in water. It does, however, dissolve in excess NH_3 because of the formation of the stable complex ion $[Ag(NH_3)_2]^+$ of Equation (2). The addition of excess HNO_3 to the clear solution causes equilibrium (2) to shift to the left, and the white precipitate of AgCl reappears. The reappearance is due to a lowering of the concentration of NH_3 owing to its reaction with H^+ to form NH_4^+.

The formation of metal complexes is often accompanied by very striking changes in color. One example with which you may be familiar is the use of solutions of $CoCl_2$ as invisible ink. What is written with this solution is not visible until the paper on which it is written is heated. The result of heating is a very legible blue color, which then slowly disappears. The phenomenon responsible for the appearance of color is shown by equilibrium (3). The pink aquo (water) complex $[Co(H_2O)_6]^{2+}$ is almost colorless when dilute, so

$$2[Co(H_2O)_6]Cl_2 \rightleftharpoons Co[CoCl_4] + 12H_2O \quad (3)$$

pink ("colorless" when dilute) — blue

that writing done with it is practically invisible. Upon the application of heat water is driven off and the blue complex $[CoCl_4]^=$ is formed. Its color is sufficiently intense that the writing can easily be read. Upon standing, water is slowly taken up from the atmosphere and the original colorless complex is regenerated, which makes the writing again invisible.

These examples serve to illustrate our point that coordination compounds are common and are frequently encountered. Until the beginning of this century the nature of these materials was not understood, and the compounds were referred to as "complex compounds." This term is still used, but fortunately no longer for the same reason. As a result of extensive research on such systems as these, our knowledge has so increased that the systems are no longer considered complicated. In fact, a knowledge of the properties of complexes is necessary to the understanding of the chemistry of the metals.

1–2 HISTORICAL DEVELOPMENT

One should keep in mind that scientific development usually comes about in a somewhat regular fashion. The collection of facts by means of many carefully designed experiments is followed by an attempt to explain and correlate all the facts with a suitable theory. It must be remembered that, unlike facts, theories can and do often change as more information becomes available. The discussions that follow in this chapter and the next may serve as good examples of how theories are modified or even at times completely discarded.

Discovery

It is difficult to state exactly when the first metal complex was discovered. Perhaps the earliest one on record is Prussian blue, $KCN \cdot Fe(CN)_2 \cdot Fe(CN)_3$, which was obtained by the artists' color maker Diesbach, in Berlin, at the beginning of the eighteenth century. However, the date usually cited is that of the discovery of hexaamminecobalt(III) chloride, $CoCl_3 \cdot 6NH_3$, by Tassaert (1798). This discovery marks the real beginning of coordination chemistry, because the existence of a compound with the unique properties of $CoCl_3 \cdot 6NH_3$ stimulated considerable interest in and research on similar systems. Although Tassaert's discovery was accidental, his realization that here was something new and different was certainly no accident, but a demonstration of his keen research ability.

Tassaert's experimental observations could not be explained on the basis of the chemical theory available at that time. It was necessary to understand how $CoCl_3$ and NH_3, each a stable compound of presumably saturated valence, could combine to make yet another very stable compound. That they could was a puzzle to chemists and a stimulus to further research, but the answer was not to be found until approximately 100 years later. During that time many such compounds were prepared and their properties studied. Several theories were proposed only to be discarded because they were inadequate to explain subsequent experimental data.

Preparation and Properties

The preparation of metal complexes generally involves the reaction between a salt and some other molecule or ion (Chapter IV).

TABLE 1-1

Compounds Named after Their Discoverers

Complex	*Name*	*Present formulation*
$Cr(SCN)_3 \cdot NH_4SCN \cdot 2NH_3$	Reinecke's salt	$NH_4[Cr(NH_3)_2(NCS)_4]$
$PtCl_2 \cdot 2NH_3$	Magnus's green salt	$[Pt(NH_3)_4][PtCl_4]$
$Co(NO_2)_3 \cdot KNO_2 \cdot 2NH_3$	Erdmann's salt	$K[Co(NH_3)_2(NO_2)_4]$
$PtCl_2 \cdot KCl \cdot C_2H_4$	Zeise's salt	$K[Pt(C_2H_4)Cl_3]$

Much of the early work was done with ammonia, and the resulting complexes were and are known as *metal ammines*. It was also found that other amines and anions such as CN^-, NO_2^-, NCS^-, and Cl^- form metal complexes. Many compounds were prepared from these anions, and at first each was named after the chemist who originally prepared it (Table 1–1). Some of these names are still used, but it soon became apparent that the system of nomenclature was not satisfactory.

Since many of the compounds are colored, the next scheme was to name compounds on the basis of color (Table 1–2). The reason for this scheme was that the colors of chloroammine complexes of cobalt(III) and chromium(III) containing the same number of ammonia molecules were found to be very nearly the same. Later the scheme was used to designate the number of ammonias without regard to color. For example, $IrCl_3{\cdot}6NH_3$ is white and not yellow as implied by the prefix *luteo*. Clearly such a system was not practical, and it had to be abandoned. The nomenclature system now used is described at the end of this chapter.

The chloroammine complexes of cobalt(III) [and those of chromium(III)] not only exhibit a spectrum of colors but also differ in the reactivity of their chlorides. For example, the addition of a solution of silver nitrate to a freshly prepared solution of $CoCl_3{\cdot}6NH_3$ results

TABLE 1–2

Compounds Named According to Their Color

Complex	*Color*	*Name*	*Present formulation*
$CoCl_3{\cdot}6NH_3$	Yellow	Luteocobaltic chloride	$[Co(NH_3)_6]Cl_3$
$CoCl_3{\cdot}5NH_3$	Purple	Purpureocobaltic chloride	$[Co(NH_3)_5Cl]Cl_2$
$CoCl_3{\cdot}4NH_3$	Green	Praseocobaltic chloride	*trans*-$[Co(NH_3)_4Cl_2]Cl$
$CoCl_3{\cdot}4NH_3$	Violet	Violeocobaltic chloride	*cis*-$[Co(NH_3)_4Cl_2]Cl$
$CoCl_3{\cdot}5NH_3{\cdot}H_2O$	Red	Roseocobaltic chloride	$[Co(NH_3)_5H_2O]Cl_3$
$IrCl_3{\cdot}6NH_3$ [a]	White	Luteoiridium chloride	$[Ir(NH_3)_6]Cl_3$

[a] This compound was called *luteo* because it contains six ammonia molecules, not because of its color (see text).

TABLE 1–3

Number of Chloride Ions Precipitated as AgCl

Complex	*Number of Cl^- ions precipitated*	*Present formulation*
$CoCl_3 \cdot 6NH_3$	3	$[Co(NH_3)_6]^{3+}$, $3Cl^-$
$CoCl_3 \cdot 5NH_3$	2	$[Co(NH_3)_5Cl]^{2+}$, $2Cl^-$
$CoCl_3 \cdot 4NH_3$	1	$[Co(NH_3)_4Cl_2]^+$, Cl^-
$IrCl_3 \cdot 3NH_3$	0	$[Ir(NH_3)_3Cl_3]$

in the immediate precipitation of all three chloride ions. The same experiment with $CoCl_3 \cdot 5NH_3$ is found to cause instant precipitation of only two chloride ions; the third chloride precipitates slowly on prolonged standing. The results of such studies are summarized in Table 1–3. These observations suggest that in $CoCl_3 \cdot 6NH_3$ and in $IrCl_3 \cdot 6NH_3$ all chlorides are identical; but in $CoCl_3 \cdot 5NH_3$ and $CoCl_3 \cdot 4NH_3$ there are two different kinds of chlorides. One type is perhaps similar to that in sodium chloride and is readily precipitated as silver chloride, whereas the other type is held more firmly and does not precipitate.

Another kind of experiment provides useful information about the number of ions present in solutions of different complexes. The greater the number of ions in a solution, the greater is the electrical conductivity of the solution. Therefore, a comparison of the conductivities of solutions containing the same concentrations of coordination compounds permits an estimate of the number of ions in each complex compound. This type of information was obtained for several series of complexes; some of the data are presented in Table 1–4. The results show that as the number of ammonia molecules in the compounds decreases, the number of ions also falls to zero and then increases again.

One other important early observation was that certain complexes exist in two different forms having the same chemical composition. Examples are the green and violet forms of $CoCl_3 \cdot 4NH_3$.

TABLE 1–4

Molar Conductivity of Platinum(IV) Complexes

Complex	*Molar conductivity, ohm^{-1}*	*Number of ions indicated*	*Present formulation*
$PtCl_4{\cdot}6NH_3$	523	5	$[Pt(NH_3)_6]^{4+}$, $4Cl^-$
$PtCl_4{\cdot}5NH_3$	404	4	$[Pt(NH_3)_5Cl]^{3+}$, $3Cl^-$
$PtCl_4{\cdot}4NH_3$	229	3	$[Pt(NH_3)_4Cl_2]^{2+}$, $2Cl^-$
$PtCl_4{\cdot}3NH_3$	97	2	$[Pt(NH_3)_3Cl_3]^+$, Cl^-
$PtCl_4{\cdot}2NH_3$	0	0	$[Pt(NH_3)_2Cl_4]$
$PtCl_4{\cdot}NH_3{\cdot}KCl$	109	2	K^+, $[Pt(NH_3)Cl_5]^-$
$PtCl_4{\cdot}2KCl$	256	3	$2K^+$, $[PtCl_6]^{2-}$

The colors of the two forms are not always so dramatically different; other physical and chemical properties also differ. For example, the α and β forms of $PtCl_2{\cdot}2NH_3$ are both a cream color, but they differ in solubility and in chemical reactivity.

It was necessary to account for all of these experimental facts with a suitable theory. Several hypotheses and theories were proposed. We shall discuss one that was used rather extensively and then proved to be wrong. We shall also discuss the coordination theory of Werner, which has withstood the test of time and provides a suitable explanation for the existence and behavior of metal complexes.

Blomstrand-Jorgensen Chain Theory

The development of a structural theory for organic compounds antedated that for coordination compounds; thus at the time people began to consider the structure of complexes, the concept of the tetravalency of carbon and the formation of carbon-carbon chains in organic compounds was already well recognized. This concept had a marked influence on the thinking of chemists of that time. No doubt it influenced Blomstrand, professor of chemistry at the University of

Lund, in Sweden, who in 1869 proposed the chain theory to explain the existence of metal complexes.

Because it was felt that elements had only one type of valence, Blomstrand and his student Jorgensen, who was later professor at the University of Copenhagen, suggested there could be only three bonds to cobalt(III) in its complexes. Therefore, a chain structure was used to account for the additional six ammonia molecules in $CoCl_3 \cdot 6NH_3$(I). The three chlorides are separated by some distance from cobalt and were therefore believed to precipitate readily as silver chloride on the addition of Ag^+. The theory represented $CoCl_3 \cdot 5NH_3$ as II. In this structure one chloride is attached directly

$$\text{Co}\begin{cases} NH_3\text{—}Cl \\ NH_3\text{—}NH_3\text{—}NH_3\text{—}NH_3\text{—}Cl \\ NH_3\text{—}Cl \end{cases} \qquad \text{Co}\begin{cases} Cl \\ NH_3\text{—}NH_3\text{—}NH_3\text{—}NH_3\text{—}Cl \\ NH_3\text{—}Cl \end{cases}$$

I II

to cobalt, and it was presumed that this is the one that does not ionize and does not precipitate instantly as silver chloride. Structure III for $CoCl_3 \cdot 4NH_3$ is also in accord with experiments that show that two chlorides are held more firmly than the third.

The next member of this series, $CoCl_3 \cdot 3NH_3$, was represented as IV. From this structure one would predict that the chlorides would

$$\text{Co}\begin{cases} Cl \\ NH_3\text{—}NH_3\text{—}NH_3\text{—}NH_3\text{—}Cl \\ Cl \end{cases} \qquad \text{Co}\begin{cases} Cl \\ NH_3\text{—}NH_3\text{—}NH_3\text{—}Cl \\ Cl \end{cases}$$

III IV

behave as they do in $CoCl_3 \cdot 4NH_3$. Professor Jorgensen, a very able experimentalist, did not succeed in preparing the cobalt compound but made instead the analogous iridium complex $IrCl_3 \cdot 3NH_3$. A solution of this compound did not conduct a current, nor did it give a precipitate upon addition of silver nitrate. Thus Jorgensen had succeeded in showing that his chain theory could not be correct.

Werner's Coordination Theory

Our present understanding of the nature of metal complexes is due to the ingenious insight of Alfred Werner, professor of chemistry in Zurich and winner of a Nobel prize in 1913. In 1893, at the age of only 26, he proposed what is now commonly referred to as *Werner's coordination theory*, and his theory has been a guiding principle in inorganic chemistry and in the concept of valence. Three of its more important postulates are:

1. Most elements exhibit two types of valence: (*a*) primary valence (——), and (*b*) secondary valence (----). In modern terminology, (*a*) corresponds to *oxidation state* and (*b*) to *coordination number*.

2. Every element tends to satisfy both its primary and secondary valence.

3. The secondary valence is directed toward fixed positions in space. [Note that this is the basis for the stereochemistry of metal complexes (Chapter III).]

With reference to these postulates let us return to the experimental facts described earlier and see how Werner's coordination theory explains them. Again it is convenient to use the chloroamminecobalt(III) complexes. According to the theory, the first member of the series, $CoCl_3 \cdot 6NH_3$, is designated as V and formulated as $[Co(NH_3)_6]Cl_3$. The primary valence or oxidation state of cobalt(III) is 3. The three chloride ions saturate the primary valence of cobalt; the ions that neutralize the charge of the metal ion use the primary valence. The secondary valence, or *coordination number* (often abbreviated CN), of Co(III) is 6. The coordination number is the number of atoms or molecules that are directly attached to the metal atom. The ammonia molecules use the secondary valence. They are said to be *coordinated* to the metal, and they are called *ligands*. Ligands (in this case ammonia) are directly attached to the atom, and they are said to be in the *coordination sphere* of the metal. Here Co(III) is already surrounded by six ammonias, so that the chloride

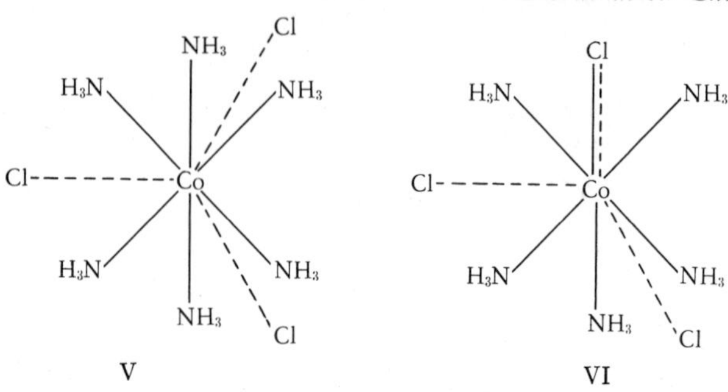

ions can not be accommodated as ligands and hence are farther from the metal ion and are not firmly bound. Thus a solution of the complex conducts current equivalent to four ions, and the chloride ions are readily precipitated by Ag^+ as silver chloride.

Returning now to the theory, one finds that Werner represented $CoCl_3 \cdot 5NH_3$ as VI. He did this in accord with postulate 2, which states that both the primary and secondary valence tend to be satisfied. In $CoCl_3 \cdot 5NH_3$ there are only five ammonia molecules to satisfy the secondary valence. Therefore, one chloride ion must serve the dual function of satisfying both a primary and a secondary valence. Werner represented the bond between such a ligand and the central metal by a combined dashed-solid line ═══. Such a chloride is not readily precipitated from solution by Ag^+. The complex cation $[Co^{(III)}(NH_3)_5Cl]^{2+}$ carries a charge of 2+ because $Co^{3+} + Cl^- = +3 - 1 = +2$. The compound $CoCl_3 \cdot 5NH_3$ is now formulated $[Co(NH_3)_5Cl]Cl_2$.

Extension of this theory to the next member of the series,

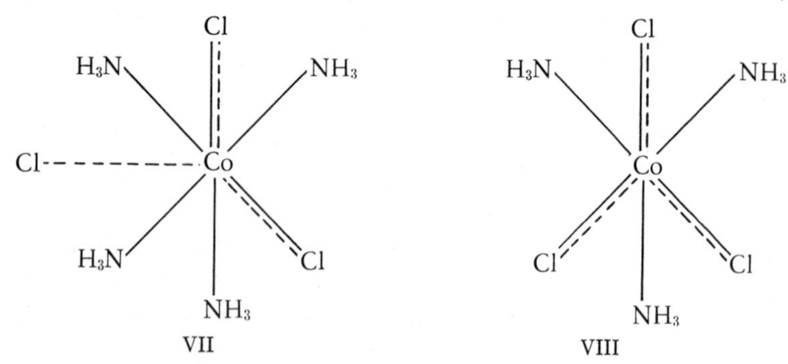

$CoCl_3 \cdot 4NH_3$, requires formula VII. Two chloride ions satisfy both primary and secondary valence; hence they are firmly held in the coordination sphere. In solution the compound therefore dissociates into two ions, Cl^- and $[Co(NH_3)_4Cl_2]^+$.

A very significant point arises with the next member of the series, $CoCl_3 \cdot 3NH_3$. Werner's theory requires that it be represented as VIII and formulated as $[Co(NH_3)_3Cl_3]$. The Werner theory predicted that the complex would not yield Cl^- ion in solution. In contrast the chain theory predicted that the compound would dissociate to give one chloride ion. The experimental results (Tables 1–3 and 1–4), when finally obtained, showed that compounds of the type $[M^{(III)}(NH_3)_3Cl_3]$ do not ionize in solution. This proved that the chain theory was wrong and supported the coordination theory.

Postulate 3 of Werner's theory deals specifically with the stereochemistry of metal complexes. Coordination theory correctly explains many of the structural features of coordination compounds. One particularly important contribution was the determination of the structure of six-coordinated complexes. *Isomers* are defined as compounds having the same formula but different structures. Before the discovery of x rays, the spatial configuration of molecules was determined by comparing the number of known isomers with the number theoretically possible for several possible structures. By this method it was possible to prove that certain structures were not correct and to obtain evidence in support of (but not proof of) a particular configuration.

Such a procedure was used successfully by Werner to demonstrate that six-coordinated complexes have an octahedral structure. The method starts with the assumption that a six-coordinated system has a structure in which the six ligands are situated at positions symmetrically equidistant from the central atom. If it is further assumed that three of the more probable structures are (1) planar, (2) a trigonal prism, and (3) octahedral (Table 1–5), then it is possible to compare the number of known isomers with the number theoretically predicted for each of these structures. Such a comparison shows that for the second and third compounds in Table 1–5 the planar and the trigonal prism structures predict there should be three isomers. Instead, complexes of these types were found to exist only in two forms, in accord with the number of isomers theoretically possible for an octahedral structure.

TABLE 1-5

Known Isomers vs. the Number Theoretically Possible for Three Different Structures

Complexes	*Number of known isomers*	*Planar*	*Trigonal prism*	*Octahedral*
MA_5B	One	One	One	One
MA_4B_2	Two	Three (1,2; 1,3; 1,4)[a]	Three (1,2; 1,4; 1,6)	Two (1,2; 1,6)
MA_3B_3	Two	Three (1,2,3; 1,2,4; 1,3,5)	Three (1,2,3; 1,2,4; 1,2,6)	Two (1,2,3; 1,2,6)

[a] Numbers indicate positions of B groups.

These results furnish negative evidence, not positive proof, that the planar and trigonal prism structures are incorrect. Failure to obtain a third isomer does not guarantee that the complexes do not have these two structures; the third isomer may be much less stable or more difficult to isolate. Since negative evidence can only indicate what may be involved, the scientist is forced to design an experiment that gives positive evidence or proof. Werner was able to prove definitely that the planar and the trigonal prism structures cannot be correct. The proof involved demonstrating that complexes of the type $[M(AA)_3]$ are optically active, as is discussed in Chapter III.

1-3 NOMENCLATURE

A comprehensive system of nomenclature of coordination compounds was not possible prior to Werner's coordination theory. As soon as it was realized that coordination compounds are either salts or nonionic species, it was possible to devise a systematic scheme for naming them. The salts were designated in the usual fashion of using a two-word name, and the nonionic compounds were given a one-word name. For example, $[Co(NH_3)_6]Cl_3$ was called hexamminecobalti chloride and $[Pt(NH_3)_2Cl_2]$ was called dichlorodiammineplatino. According to this scheme the suffixes -a, -o, -i, and -e were used to designate the +1, +2, +3, and +4 oxidation states of the metal. This method has now been replaced by the Stock system of using Roman numerals in parentheses to indicate oxidation states. Thus $[Co(NH_3)_6]Cl_3$ is called hexaamminecobalt(III) chloride and $[Pt(NH_3)_2Cl_2]$ is dichlorodiammineplatinum(II). The nomenclature system outlined below and that used throughout this text is the one recommended by the Inorganic Nomenclature Committee of the International Union of Pure and Applied Chemistry.

Order of Listing Ions

The cation is named first, and then the anion. This is the usual practice when naming a salt.

$NaCl$	Sodium Chloride
$[Cr(NH_3)_6](NO_3)_3$	hexaamminechromium(III) nitrate
$K_2[PtCl_6]$	potassium hexachloroplatinate(IV)

Nonionic Complexes

Nonionic or molecular complexes are given a one-word name.

$[Co(NH_3)_3(NO_2)_3]$	trinitrotriamminecobalt(III)
$[Cu(CH_3COCHCOCH_3)_2]$	bis(acetylacetonato)copper(II)

$CH_3COCH_2COCH_3$ = acetylacetone

Names of Ligands

Neutral ligands are named as the molecule; negative ligands end in -o; and positive ligands (of rare occurrence) end in -ium.

$NH_2CH_2CH_2NH_2$	ethylenediamine
$(C_6H_5)_3P$	triphenylphosphine
Cl^-	chloro
CH_3COO^-	acetato
$NH_2NH_3^+$	hydrazinium

Two exceptions to this rule are water and ammonia.

H_2O aquo
NH_3 ammine (Note the spelling with two m's; this applies only to NH_3; other amines are spelled with the usual one m.)

Order of Ligands

The ligands in a complex are named in the order (1) negative, (2) neutral, and (3) positive, without separation by hyphens. Within each of these categories the groups are listed in order of increasing complexity.

$[Pt(NH_3)_4(NO_2)Cl]SO_4$ chloronitrotetraammineplatinum(IV) sulfate
$NH_4[Cr(NH_3)_2(NCS)_4]$ ammonium tetrathiocyanatodiamminechromate(III)

Numerical Prefixes

The prefixes di-, tri-, tetra-, etc., are used before such simple expressions as bromo, nitro, and oxalato. Prefixes bis-, tris-, tetrakis-, etc., are used before complex names (chiefly expressions containing the prefixes mono-, di-, tri-, etc., in the ligand name itself) such as ethylenediamine and trialkylphosphine.

$K_3[Al(C_2O_4)_3]$ potassium trioxalatoaluminate(III)
$[Co(en)_2Cl_2]_2SO_4$ dichlorobis(ethylenediamine)cobalt(III) sulfate
en = $NH_2CH_2CH_2NH_2$ = ethylenediamine

Termination of Names

The ending for anionic complexes is -ate; alternatively, -ic if named as an acid. For cationic and neutral complexes the name of the metal is used without any characteristic ending.

$Ca_2[Fe(CN)_6]$	calcium hexacyanoferrate(II)
$[Fe(H_2O)_6]SO_4$	hexaaquoiron(II) sulfate
$[Ni(DMG)_2]$	bis(dimethylglyoximato)nickel(II)

DMG = 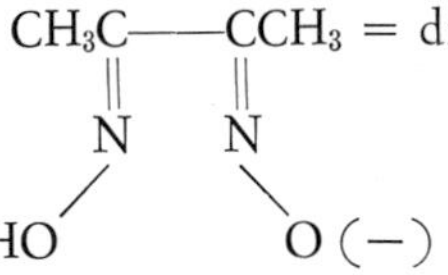 = dimethylglyoximate ion

Oxidation States

The oxidation state of the central atom is designated by a Roman numeral in parentheses at the end of the name of the complex, without a space between the two. For a negative oxidation state a minus sign is used before the Roman numeral, and 0 is used for zero.

$Na[Co(CO)_4]$	sodium tetracarbonylcobaltate(−I)
$K_4[Ni(CN)_4]$	potassium tetracyanonickelate(0)

Bridging Groups

Ligands that bridge two centers of coordination are preceded by the Greek letter μ, which is repeated before the name of each different kind of bridging group.

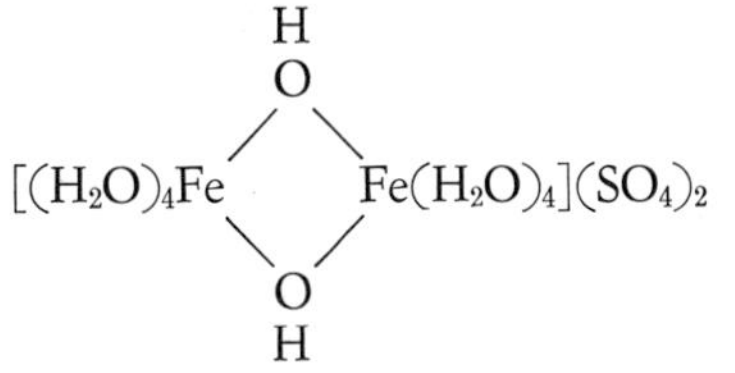

octaaquo-μ-dihydroxo-diiron(III) sulfate

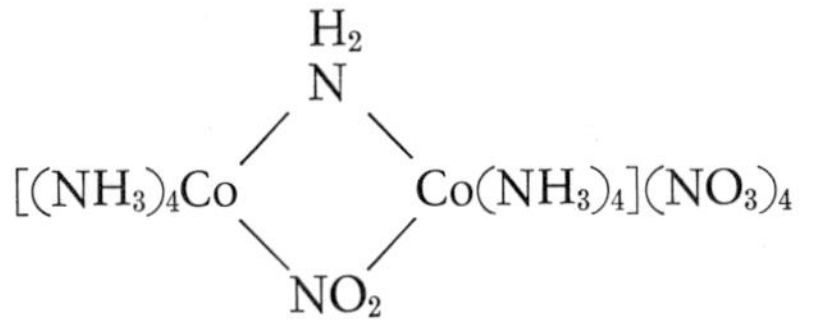

octaammine-μ-amido-μ-nitrodicobalt(III) nitrate

Point of Attachment

Whenever necessary, the point of attachment of a ligand is designated by placing the symbol (in italics) of the element attached after the name of the group, with separation by hyphen.

$(NH_4)_3[Cr(NCS)_6]$	ammonium hexathiocyanato-*N*-chromate(III)
$(NH_4)_2[Pt(SCN)_6]$	ammonium hexathiocyanato-*S*-platinate(IV)

For thiocyanate and nitrite ions the following optional names may be used:

—SCN^-, thiocyanato	—NO_2^-, nitro
—NCS^-, isothiocyanato	—ONO^-, nitrito

Geometrical Isomers

Geometrical and optical isomerism are discussed in Sections 3–3 and 3–4. The naming of isomeric materials is included at this point for the sake of completeness. Geometrical isomers are generally named by the use of the terms *cis* to designate adjacent (90° apart) positions and *trans* for opposite (180° apart) positions. It is occasionally necessary to use a number system to designate the position of each ligand. For square planar complexes, groups 1–3 and 2–4 are in

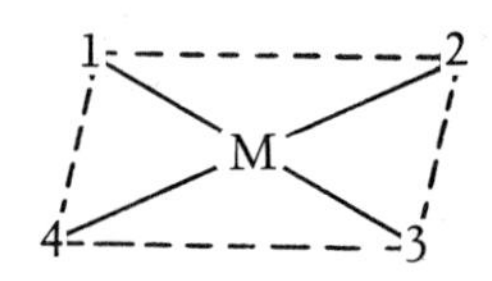

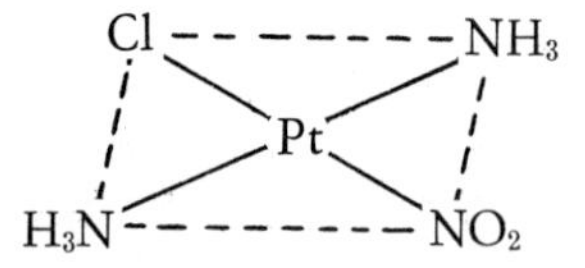

trans-chloronitrodiammineplatinum(II)

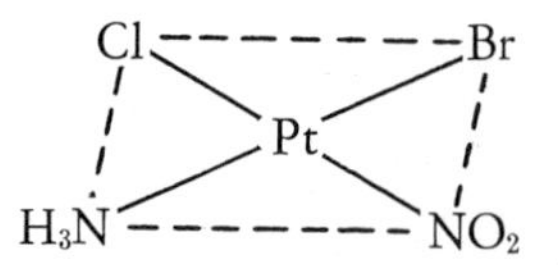

1-chloro-3-nitrobromo ammineplatinum(II) ion

trans positions. Note that only two of the *trans* positions need be numbered in the name of the second complex. This is due to the fact that in a square complex the other two ligands must then be in *trans* positions. Since positions 2 and 4 are equivalent, these numbers need not be mentioned. Simple models are most helpful in visualizing these complexes.

The number system for octahedral complexes has the *trans* positions numbered 1–6, 2–4, and 3–5. An optional name for the last compound is *trans*-chloronitro-*trans*-bromoiodoamminepyridineplatinum(IV).

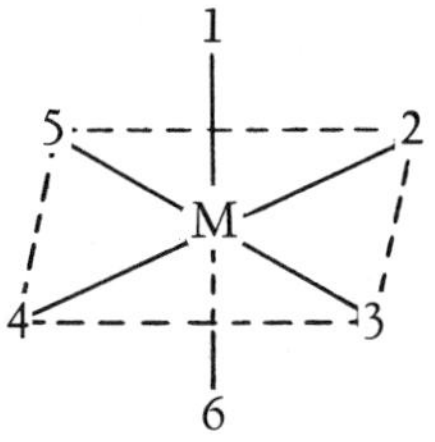

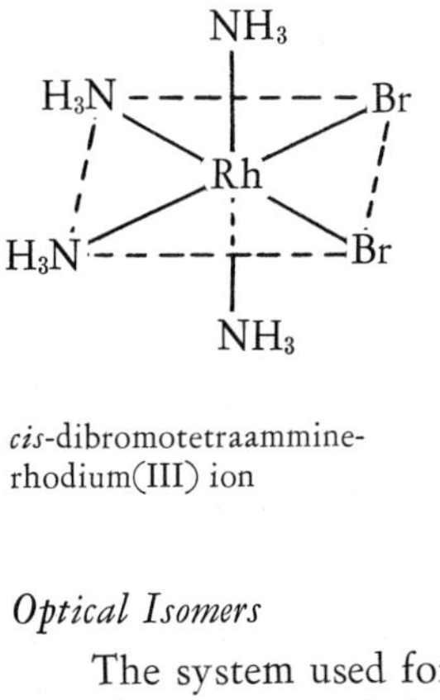

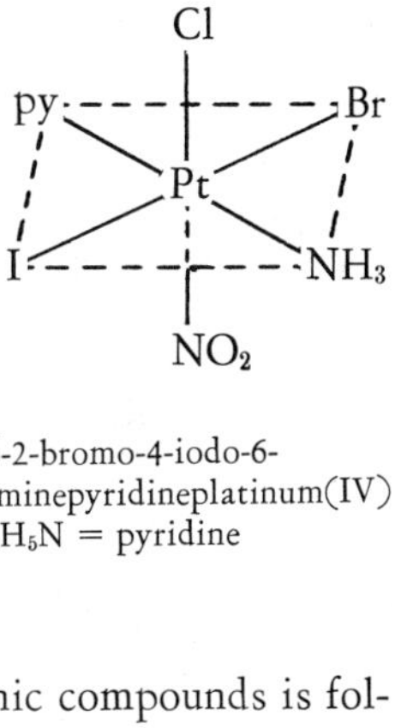

cis-dibromotetraammine-rhodium(III) ion

1-chloro-2-bromo-4-iodo-6-nitroamminepyridineplatinum(IV)
py = C_5H_5N = pyridine

Optical Isomers

The system used for optically active organic compounds is followed. *Dextro*- and *levo*rotatory compounds are respectively designated either by (+) or (−) or by *d* or *l*.

(+), or *d*−$K_3[Ir(C_2O_4)_3]$ potassium(+), or *d*−trioxalatoiridate(III)

Abbreviations

Simple abbreviations are customarily used for complicated molecules in coordination compounds. Unfortunately, there is still no definite agreement on the abbreviation to use for a particular ligand. Some of the abbreviations that are commonly used and those used in this book are given in Table 1–6.

Miscellaneous Terminology

It is convenient to include here the definition or description of some terms not yet introduced. Ethylenediamine (en) occupies two coordination positions and thus behaves somewhat as if it were two molecules of ammonia tied together. Other molecules have the capacity to attach to the central atom at even more than two positions; for example, dien and EDTA may be attached to three and six positions, respectively (Table 1–6). Such a group is called a *multidentate* or a *chelate* ligand. The correct usage of the adjective "chelate" is best illustrated by a specific example. The salt $[Cu(en)_2]SO_4$, IX, is designated as a *chelate compound*, the cation as a

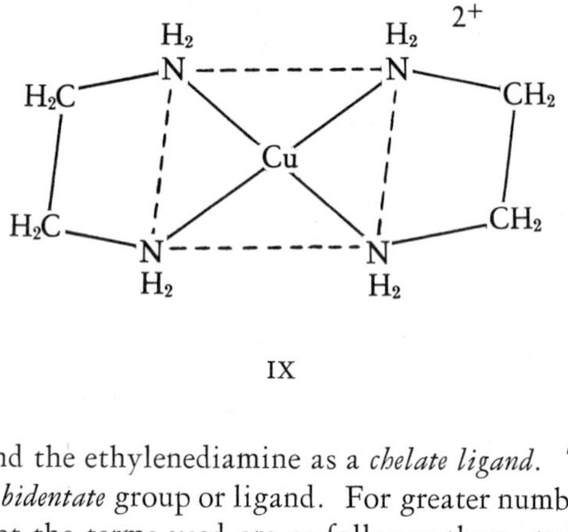

IX

chelate ion, and the ethylenediamine as a *chelate ligand*. The latter is also called a *bidentate* group or ligand. For greater numbers of points of attachment the terms used are as follows: three, *terdentate;* four, *quadridentate;* five, *quinquidentate;* and six, *sexadentate*.

Whenever a ligand atom is attached to two metal ions, it is called a *bridging group* (see Bridging Groups, Sec. 1–3). The resulting complex is often called a "polynuclear complex," but a better term, also used, is *bridged complex*. This is preferred because the prefix "poly" usually denotes a high molecular weight, whereas these substances are often only dimers or trimers.

TABLE 1–6

Symbols Used for Some Ligands

Symbol	*Ligand Name*	*Formula*
en	ethylenediamine	$\ddot{N}H_2CH_2CH_2\ddot{N}H_2$
py	pyridine	N:
pn	propylenediamine	$\ddot{N}H_2CH_2CH(CH_3)\ddot{N}H_2$
dien	diethylenetriamine	$\ddot{N}H_2CH_2CH_2\ddot{N}HCH_2CH_2\ddot{N}H_2$
trien	triethylenetetramine	$\ddot{N}H_2CH_2CH_2\ddot{N}HCH_2CH_2\ddot{N}HCH_2CH_2\ddot{N}H_2$
bipy	2,2′-bipyridine	N N
phen	1,10-phenanthroline	N N
EDTA	ethylenediaminetetraacetato	$^{-}$:OOCCH$_2$, CH$_2$COO:$^{-}$, $\ddot{N}CH_2CH_2\ddot{N}$, $^{-}$:OOCCH$_2$, CH$_2$COO:$^{-}$
DMG	dimethylglyoximato	CH_3C—CCH_3 (‖ N: :N) HO O$^-$
gly	glycinato	$:NH_2CH_2COO:^-$
acac	acetylacetonato	$CH_3CCH = CCH_3$ (‖ $\ddot{O}$, | $\ddot{O}(-)$)

PROBLEMS

1. The compound $CoCl_3 \cdot 2en$ (en = $NH_2CH_2CH_2NH_2$) contains only one chloride ion that will be precipitated immediately upon addition of silver ion. (*a*) Draw the structure of this compound on the basis of the Blomstrand-Jørgensen chain theory of bonding. (*b*) Draw the structure on the basis of Werner's coordination theory. (*c*) Explain how each theory accounts for there being only one ionic chloride. (*d*) Explain why the chain theory cannot account for the stereochemistry of the compound.

2. Combinations of Co(III), NH_3, NO_2^-, and K^+ can result in the formation of a series of seven coordination compounds, one of which is $[Co(NH_3)_6](NO_2)_3$. (*a*) Write the formulas for the other six members of the series. (*b*) Name each compound. (*c*) Indicate the complexes that should form geometric isomers (Section 3–3).

3. (*a*) Name each of the following compounds:

$[Co(NH_3)_6]_2(SO_4)_3$

$K_4[(C_2O_4)_2Co(OH)_2Co(C_2O_4)_2]$ (two bridging OH groups between the Co atoms)

Pt with H_3CNH_2, Cl, Br, NH_3 (square planar; H_3CNH_2 trans to NH_3, Cl trans to Br)

$[Co(NH_3)_4(NCS)Cl]NO_3$

$[Pt(en)Cl_4]$

$NH_4[Cr(NH_3)_2(NCS)_4]$

$Na_2[Ni(EDTA)]$

$K_4[Ni(CN)_4]$

Pt with $H_2NCH_2CH_2NH_2$ (chelating), two NO_2, and two axial Cl (octahedral)

(*b*) Write the formula for each of the following compounds:
dibromotetraammineruthenium(III) nitrate
chloroaquobis(ethylenediamine)rhodium(III) chloride
calcium dioxalatodiamminecobaltate(III)
sodium tetrahydroxoaluminate(III)
cesium fluorotrichloroiodate(III)
octaammine-μ-amido-μ-hydroxodicobalt(III) sulfate
trans-diglycinatopalladium(II)
sodium dithiosulfato-*S*-argentate(I)

4. Solid $CrCl_3 \cdot 6H_2O$ may be either $[Cr(H_2O)_6]Cl_3$, $[Cr(H_2O)_5Cl]Cl_2 \cdot H_2O$, or $[Cr(H_2O)_4Cl_2]Cl \cdot 2H_2O$. By making use of an ion-exchange column, it is possible to determine which of these three formulas is correct.

A solution containing 0.319 g of $CrCl_3 \cdot 6H_2O$ was passed through a cation-exchange resin in the acid form, and the acid liberated was titrated with a standard solution of NaOH. This required 28.5 cc of 0.125 *M* NaOH. Determine the correct formula of the Cr(III) complex.

REFERENCES

J. C. Bailar, Jr. (ed.), *The Chemistry of Coordination Compounds*, Reinhold, New York, 1956.

A. A. Grinberg in D. H. Busch and R. F. Trimble, Jr. (eds.), *The Chemistry of Complex Compounds*, Addison-Wesley, Reading, Mass., 1962.

"Nomenclature of inorganic chemistry," *J. Am. Chem. Soc.*, **82,** 5523 (1960).

Chapters dealing with coordination chemistry are found in almost all textbooks of inorganic chemistry. The following are recommended:

T. Moeller, *Inorganic Chemistry*, Wiley-Interscience, New York, 1952.

J. Kleinberg, W. J. Argersinger, Jr., and E. Griswold, *Inorganic Chemistry*, Heath, Boston, 1960.

F. A. Cotton and G. Wilkinson, *Advanced Inorganic Chemistry*, Wiley-Interscience, New York, 1962.

II

The Coordinate Bond

Werner's coordination theory, with its concept of secondary valence, provides an adequate explanation for the existence of such complexes as $[Co(NH_3)_6]Cl_3$. The properties and stereochemistry of these complexes are also explained by the theory, which remains the real foundation of coordination chemistry. Since Werner's work antedated by about twenty years our present electronic concept of the atom, the theory was not able to describe in modern terms the nature of the secondary valence, or as it is now called, the *coordinate bond*. Three theories are currently used to describe the nature of the bonding in metal complexes. These theories are (1) the valence bond theory (VBT), (2) the electrostatic crystal field theory (CFT), and (3) the molecular orbital theory (MOT). We shall first describe the contributions of G. N. Lewis and N. V. Sidgwick to the theory of chemical bonding.

2-1 THE ELECTRON-PAIR BOND

In 1916 G. N. Lewis, professor of chemistry at the University of California in Berkeley, postulated that a bond between two atoms A and B can arise by their sharing a pair of electrons. Each atom usually contributes one electron. This electron pair bond is called a *covalent bond.* On this basis he pictured the molecules, CH_4 and NH_3 as

$$\begin{matrix} & H & \\ & \cdot\cdot & \\ H: & C & :H \\ & \cdot\cdot & \\ & H & \end{matrix} \qquad \text{and} \qquad \begin{matrix} & H & \\ & \cdot\cdot & \\ : & N & :H \\ & \cdot\cdot & \\ & H & \end{matrix}$$

respectively. This type of representation is now referred to as the *Lewis diagram* of a molecule.

An examination of the Lewis diagrams shows CH_4 and NH_3 to be similar in that there are two electrons adjacent to each hydrogen, whereas carbon and nitrogen are each associated with eight electrons. A significant and important difference is that one electron pair on nitrogen is not shared by a hydrogen. This permits the ammonia molecule to react in such a way as to share the "free" electron pair with some other atom. The resulting bond is also an electron pair or covalent bond; but because both electrons are furnished by the nitrogen, the bond is sometimes called a *coordinate covalent bond.*

The reaction of ammonia with acids to form ammonium salts (1) produces a coordinate covalent bond. The four N—H bonds in NH_4^+ are nonetheless equivalent. This indicates that the distinction between coordinate and normal covalent bonds has little meaning. Ammonia may also share its free electron pair with substances other

$$H^+ + \begin{matrix} & H & \\ & \cdot\cdot & \\ : & N & :H \\ & \cdot\cdot & \\ & H & \end{matrix} \rightarrow \left[\begin{matrix} & H & \\ & \cdot\cdot & \\ H: & N & :H \\ & \cdot\cdot & \\ & H & \end{matrix}\right]^+ \tag{1}$$

than hydrogen ion. When a metal ion takes the place of hydrogen ion, a metal ammine complex is formed, (2), (3), and (4). Since these reactions are generally carried out in aqueous solution, it is

$$Ag^+ + \begin{matrix} & H & \\ & \cdot\cdot & \\ : & N & :H \\ & \cdot\cdot & \\ & H & \end{matrix} \rightarrow \left[\begin{matrix} & H & \\ & \cdot\cdot & \\ Ag: & N & :H \\ & \cdot\cdot & \\ & H & \end{matrix}\right]^+ \xrightarrow{NH_3} \left[\begin{matrix} & H & & H & \\ & \cdot\cdot & & \cdot\cdot & \\ H: & N & :Ag: & N & :H \\ & \cdot\cdot & & \cdot\cdot & \\ & H & & H & \end{matrix}\right]^+ \tag{2}$$

$$Cu^{2+} + 4{:}NH_3 \rightarrow \left[\begin{array}{c} NH_3 \\ \cdot\cdot \\ H_3N{:}Cu{:}NH_3 \\ \cdot\cdot \\ NH_3 \end{array}\right]^{2+} \tag{3}$$

$$Ni^{2+} + 6{:}NH_3 \rightarrow \left[\begin{array}{ccc} & NH_3 & \\ H_3N & \cdot\!\cdot\!\cdot & NH_3 \\ & {:}Ni{:} & \\ H_3N & \cdot\!\cdot\!\cdot & NH_3 \\ & NH_3 & \end{array}\right]^{2+} \tag{4}$$

more correct to indicate that the ions initially present are aquo complexes and that coordinated water is replaced by ammonia (5–8).

$$[H{:}OH_2]^+ + {:}NH_3 \rightleftharpoons [H{:}NH_3]^+ + H_2O \tag{5}$$

$$[Ag({:}OH_2)_2]^+ + 2{:}NH_3 \rightleftharpoons [Ag({:}NH_3)_2]^+ + 2H_2O \tag{6}$$

$$[Cu({:}OH_2)_4]^{2+} + 4{:}NH_3 \rightleftharpoons [Cu({:}NH_3)_4]^{2+} + 4H_2O \tag{7}$$

$$[Ni({:}OH_2)_6]^{2+} + 6{:}NH_3 \rightleftharpoons [Ni({:}NH_3)_6]^{2+} + 6H_2O \tag{8}$$

These reactions are Lewis *acid-base reactions*. The *Lewis theory of acids and bases* defines an *acid* as a substance capable of accepting a pair of electrons and a *base* as a substance that donates a pair of electrons. The terms *acceptor* and *donor* are sometimes used for acid and base, respectively. An acid-base reaction results in the formation of a coordinate bond and a coordination compound (9). The Lewis

$$\begin{array}{ccccc} A & + & {:}B & \rightarrow & A{:}B \\ \text{acid} & & \text{base} & & \text{coordination} \\ \text{(acceptor)} & & \text{(donor)} & & \text{compound} \end{array} \tag{9}$$

theory is more general than the acid-base theory of Arrhenius, which defines an acid as a substance that yields hydrogen ions and a base as a substance that provides hydroxide ions. Arrhenius acids and bases are also Lewis acids and bases, as shown by the neutralization reaction (10).

$$H^+ + {:}OH^- \rightarrow H{:}OH \tag{10}$$

The Lewis acid-base concept classifies metal ions as acids. Furthermore, compounds such as BF_3, $AlCl_3$, SO_3, and SiF_4, which can accept electron pairs, are also acids. Compounds of the type F_3BNH_3 and $C_5H_5NSO_3$ are usually called *addition compounds*, but they are also examples of coordination compounds.

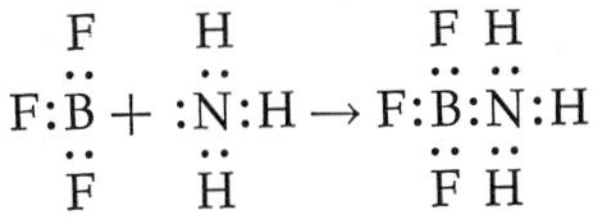

$$\begin{matrix} & \mathrm{F} & & \mathrm{H} & & \mathrm{F}\ \mathrm{H} \\ & \cdot\cdot & & \cdot\cdot & & \cdot\cdot\ \cdot\cdot \\ \mathrm{F}: & \mathrm{B} & + : & \mathrm{N} & :\mathrm{H} \rightarrow \mathrm{F}: & \mathrm{B}:\mathrm{N}:\mathrm{H} \\ & \cdot\cdot & & \cdot\cdot & & \cdot\cdot\ \cdot\cdot \\ & \mathrm{F} & & \mathrm{H} & & \mathrm{F}\ \mathrm{H} \end{matrix} \tag{11}$$

$$AlCl_3 + :Cl^- \rightarrow [AlCl_4]^- \tag{12}$$

$$SO_3 + C_5H_5N: \rightarrow C_5H_5N:SO_3 \tag{13}$$

$$SiF_4 + 2:F^- \rightarrow [SiF_6]^{2-} \tag{14}$$

Ligands share electron pairs with metals; thus ligands are Lewis bases. Some examples are the molecules $H_2O:$, $:NH_3$, $(C_2H_5)_3P:$, $:CO$, and $:NH_2CH_2CH_2\ddot{N}H_2$ and ions such as

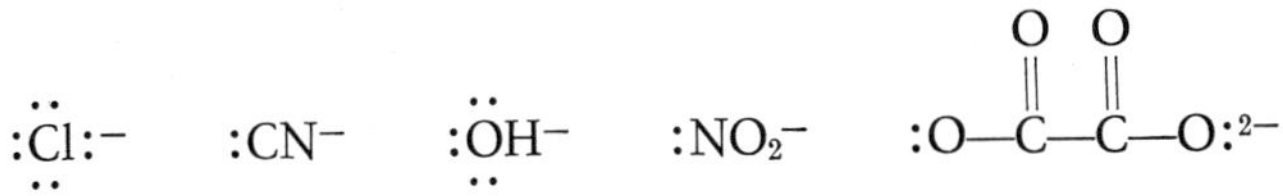

$$:\ddot{\underset{\cdot\cdot}{Cl}}:^- \qquad :CN^- \qquad :\ddot{\underset{\cdot\cdot}{O}}H^- \qquad :NO_2^- \qquad :O{-}\overset{\overset{O}{\parallel}}{C}{-}\overset{\overset{O}{\parallel}}{C}{-}O:^{2-}$$

and

$$\begin{matrix} :OOCCH_2 & & & & CH_2COO:^{4-} \\ & \diagdown & & \diagup & \\ & & \underset{\cdot\cdot}{N}CH_2CH_2\underset{\cdot\cdot}{N} & & \\ & \diagup & & \diagdown & \\ :OOCCH_2 & & & & CH_2COO: \end{matrix}$$

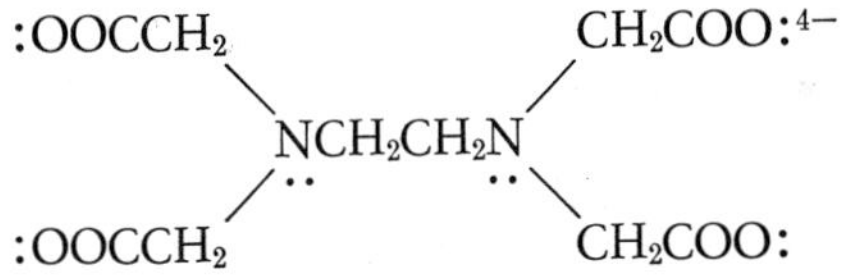

It is apparent why en and EDTA (Table 1–6) are able to function as bidentate and hexadentate ligands, respectively. Similarly, one can see that a ligand atom that contains more than one pair of free electrons may serve as a bridging atom (I).

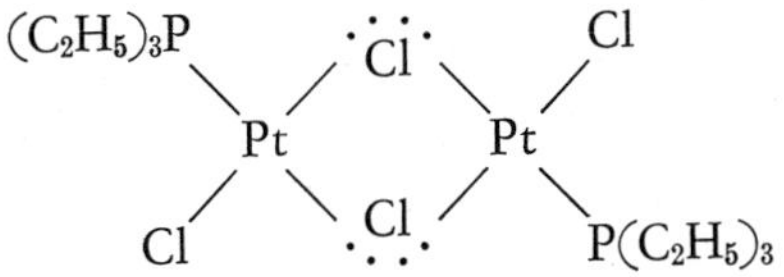

$$\begin{matrix} (C_2H_5)_3P & & \dot{\ddot{Cl}}\cdot & & Cl \\ & \diagdown\ \diagup & & \diagdown\ \diagup & \\ & Pt & & Pt & \\ & \diagup\ \diagdown & & \diagup\ \diagdown & \\ Cl & & \cdot\underset{\cdot\cdot}{Cl}\cdot & & P(C_2H_5)_3 \end{matrix}$$

I

2–2 THE CONCEPT OF EFFECTIVE ATOMIC NUMBER

The rare-gas elements (He, Ne, Ar, Kr, Xe, and Rn) are very unreactive; only very recently have compounds of these elements been prepared. It has long been recognized that compounds in which each atom can, by electron sharing with other atoms, gather around itself a number of electrons equal to that found in a rare gas also tend to be very stable. Professor N. V. Sidgwick of Oxford University applied this observation to metal complexes. He postulated that the central metal would surround itself with sufficient ligands that the total number of electrons around the metal would be the same as that in a rare gas. The number of electrons surrounding the coordinated metal is called its *effective atomic number*, which is given the symbol *EAN*. For example, the EAN of Co(III) in $[Co(NH_3)_6]^{3+}$ is readily calculated as follows:

Co atomic number 27, has 27 electrons
Co(III) $27 - 3 = 24$ electrons
6(:NH_3) $2 \times 6 = 12$ electrons[1]
EAN of Co(III) in $[Co(NH_3)_6]^{3+} = 24 + 12 = 36$ electrons

Similarly determined EAN values for other metal complexes in many cases equal the atomic numbers of rare gases. There are, however, many exceptions to this rule; examples are $[Ag(NH_3)_2]^+$ and $[Ni(en)_3]^{2+}$, with EAN values of 50 and 38, respectively. This is unfortunate; for if it were true that the EAN of the central metal always exactly equaled the atomic number of a rare gas, then it would be possible to estimate the coordination number of metal ions.

One class of compounds that frequently does obey the EAN rule comprises the metal carbonyls and their derivatives. By using the rule it is possible to predict accurately the CN of the simplest carbonyls and also predict whether the compounds can exist as monomers. For example, the EAN is 36 for the metals in the compounds $Ni(CO)_4$, $Fe(CO)_5$, $Fe(CO)_4Cl_2$, $Mn(CO)_5Br$, $CoNO(CO)_3$, and $Fe(NO)_2(CO)_2$. To estimate the EAN in these systems, it is con-

[1] Each ligand is considered to share two electrons with the central metal atom.

venient to say that, CO, Cl^-, and Br^- contribute two electrons and NO, three electrons. Manganese carbonyl has the formula $(CO)_5Mn—Mn(CO)_5$. This is the simplest formula possible if each Mn is to have an EAN of 36.

Electrons from each Mn	=	25
Electrons from 5(:CO)	=	10
Electron from Mn—Mn bond	=	1
		36

A manganese atom can gain one electron by forming a bond with another Mn atom. Each metal atom donates one electron to the bond, and each shares the two electrons.

2–3 ELECTRONIC STRUCTURE OF THE ATOM

Before continuing the discussion of bonding theory it is necessary to review briefly the electronic structure of the atom. Recall that electrons in atoms are believed to occupy successive energy levels. A maximum of 2 electrons can occupy the first level, 8 the second, 18 the third, and 32 the fourth. The principal energy levels, 1 to 7, are divided into sublevels, *s*, *p*, *d*, and *f;* successive electrons are placed in the unfilled sublevel of lowest energy. In all subsequent discussions electrons will be placed in the level of lowest energy.

By examination of the energy level diagram (Figure 2–1) it can be seen that in each level the *s* sublevel is of lower energy than the *p*, the *p* is lower than the *d*, and the *d* is lower than the *f*. The diagram also indicates that the 3*d* sublevel is of higher energy than the 4*s* and that the 4*f* is of higher energy than the 6*s;* thus sublevels of one principal level may be of higher energy than low-energy sublevels of subsequent principal levels.

Whereas the energies of sublevels of a given level always lie in the order $s < p < d < f$, the relative energies of sublevels in different levels are influenced by the surroundings of the atom and are strongly influenced by the atomic number of the atom being considered. Therefore, in the potassium atom the 3*d* sublevel is of

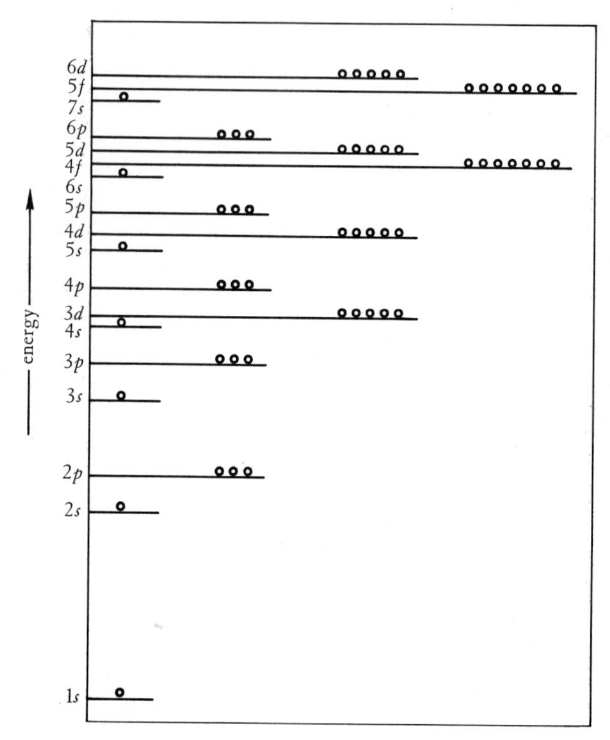

Figure 2–1 Energy level diagram for the orbitals of a light atom. The relative energies of the orbitals are accurate, but absolute energies are distorted in this diagram.

higher energy than the 4*s*, in scandium the 3*d* and 4*s* sublevels are of comparable energy, and in zinc the 4*s* sublevel is of higher energy than the 3*d*. The assignment of electronic configurations to atoms can only be approximated with the diagram in Figure 2–1. The experimentally observed configurations are presented in the periodic table at the beginning of this book.

The small circles in Figure 2–1 represent *orbitals*. The number of orbitals for each of the sublevels are *s*,1; *p*,3; *d*,5; and *f*,7. Each orbital can contain a maximum of 2 electrons, so that the maximum number of electrons for each sublevel is *s*,2; *p*,6; *d*,10; and *f*,14.

Electrons fill each sublevel in accordance with *Hund's rule*, which states that electrons in orbitals of one sublevel tend to have the same spin. This means that electrons add to empty orbitals so long as they are available. This is reasonable, since electrons repel each other and would therefore prefer to be in separate orbitals (as far from one another as possible). The electronic structures of N, Ti, and Mn can be designated as shown in Figure 2–2. Electrons in the p sublevel of N and in the d sublevel of Ti and Mn are unpaired. It is not necessary to list all of the sublevels as is done in the figure. Only the electrons beyond those of the preceding rare gas (the *valence electrons*) are generally shown, since they are the ones involved in chemical bonding. One final point to note is that it is more convenient for later use to list the $3d$ sublevel before the $4s$, the $4d$ and $4f$ before the $5s$, etc.

Having briefly discussed the electronic structures of atoms, it is now necessary to consider the electronic structures of ions. In general, in the formation of positive ions electrons are lost from the highest-energy occupied orbitals of the atom. In the case of the transition metals the highest-energy electrons are outer s electrons, and hence these are lost first. Therefore, the electronic structures of Ti^{3+} and Mn^{2+} can be represented as shown in Figure 2–3.

Next it is necessary to be familiar with the shapes of these orbitals. By the "shape of an orbital" one means the shape of the region in space in which one is most likely to find an electron residing in

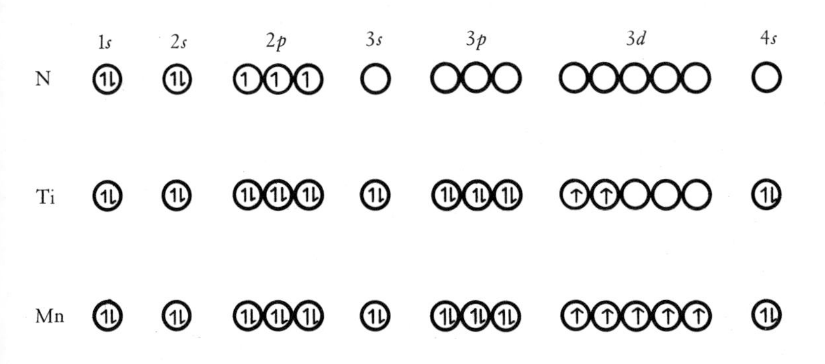

Figure 2–2 The electronic structures of N, Ti, and Mn.

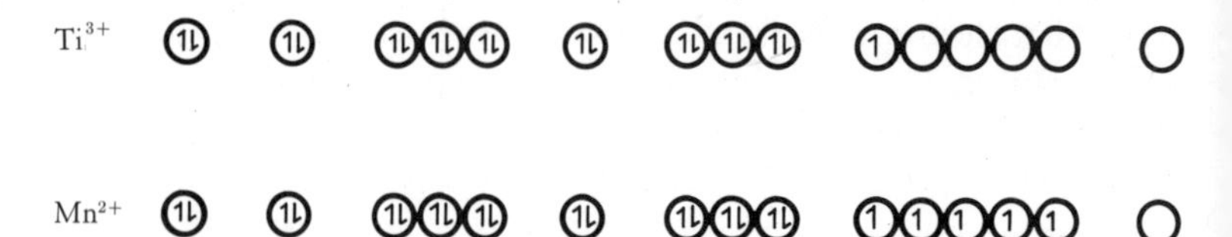

Figure 2–3 The electronic structures of Ti^{3+} and Mn^{2+}.

that orbital. We shall limit ourselves to the *s*, *p*, and *d* orbitals, the ones most commonly involved in bond formation. The *f* orbitals may be used by the inner transition elements (the rare earths and the actinides). The *s* orbital has spherical symmetry (Figure 2–4); the *p* orbitals have a dumbbell shape and each orbital is oriented along one of the three cartesian axes. The p_x orbital is oriented along the

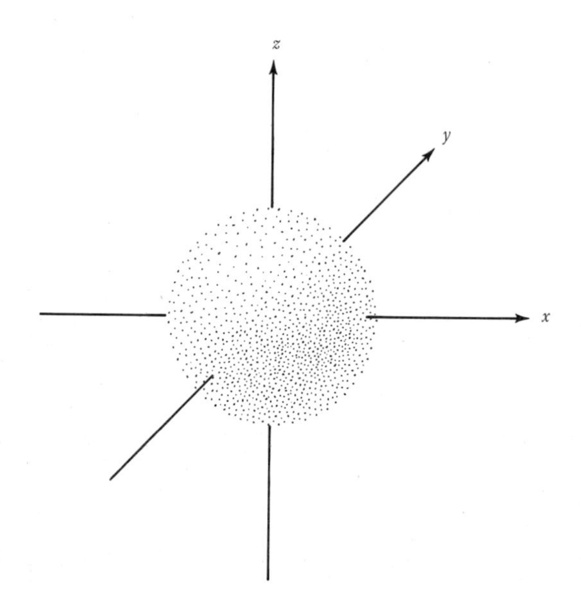

Figure 2–4 The spatial configuration of an *s* orbital.

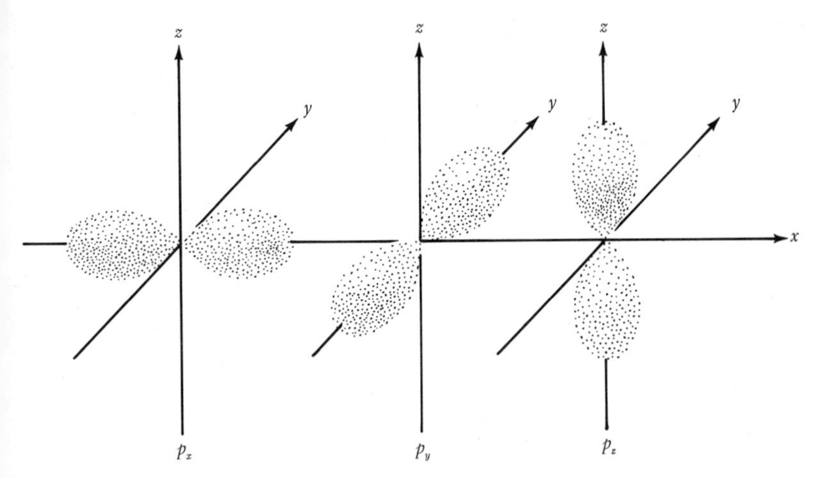

Figure 2–5 The spatial configurations of *p* orbitals.

x axis, the p_y along the y axis, and the p_z along the z axis (Figure 2–5).

Four of the d orbitals have a cloverleaf shape, and one has a dumbbell shape with a sort of sausage wrapped around the center. Three of the cloverleaf orbitals, d_{xy}, d_{xz}, and d_{yz}, are oriented in the xy, xz, and yz planes, respectively, with their lobes situated between the two axes; the other cloverleaf orbital, $d_{x^2-y^2}$, is oriented in the xy plane and has its lobes along the x and y axes (Figure 2–6). The unique dumbbell-shaped orbital, d_{z^2}, is oriented along the z axis. To understand the theories of bonding in metal complexes, it is absolutely essential to maintain a mental picture of the three-dimensional shapes of these orbitals.

2–4 VALENCE BOND THEORY

The valence bond theory was developed by Prof. Linus Pauling, of the California Institute of Technology, and made conveniently available to chemists in his excellent book, *The Nature of the Chemical Bond*, published in 1940, 1948, and 1960. Except for the late Marie

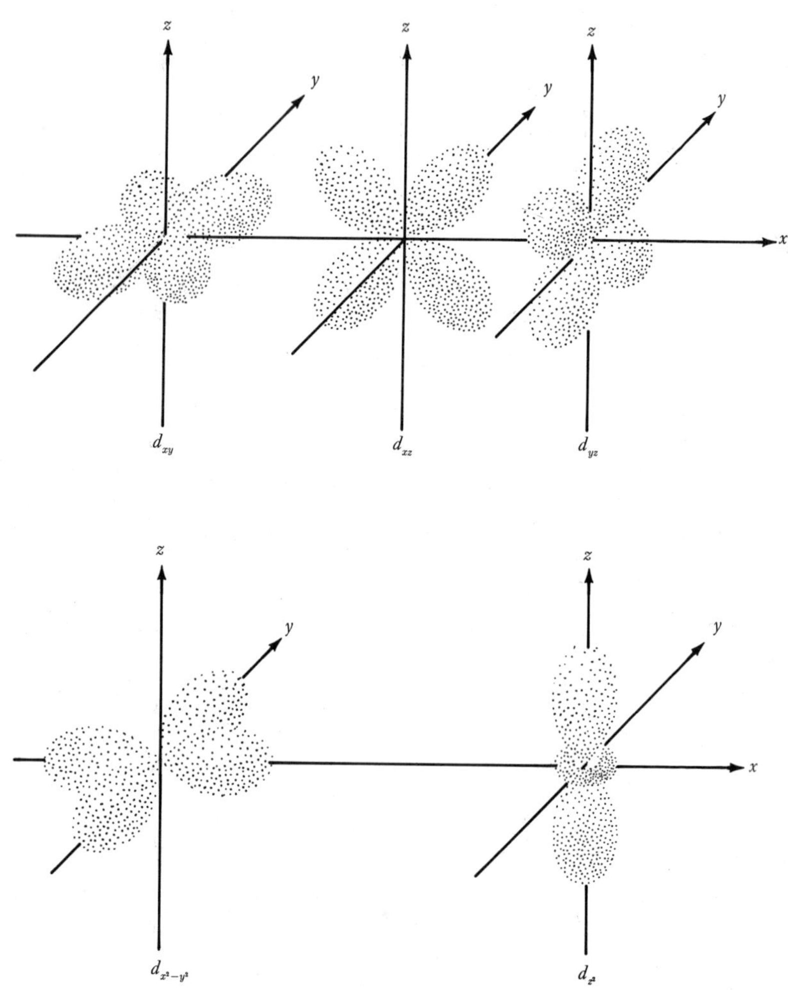

Figure 2–6 The spatial configurations of the *d* orbitals.

Curie, Professor Pauling is the only person to have been awarded two Nobel prizes. He obtained the Nobel prize in chemistry in 1954 and the Nobel peace prize in 1962. Pauling's ideas have had an important impact on all areas of chemistry; his valence bond theory has aided coordination chemists and has been extensively used. It can account reasonably well for the structure and magnetic properties of metal complexes. Extensions of the theory will account for other properties of coordination compounds such as absorption spectra, but other theories seem to do this in a more simple fashion. Therefore, in recent years coordination chemists have favored the crystal field, ligand field, and molecular orbital theories. Since space is limited, we shall concern ourselves only with the latter theories.

It may, however, be useful to show the valence bond theory representations of the complexes $[CoF_6]^{3-}$ and $[Co(NH_3)_6]^{3+}$, which can then be compared with representations of the crystal field and molecular orbital theories to be discussed later. First it is necessary to know that $[CoF_6]^{3-}$ contains four unpaired electrons, whereas $[Co(NH_3)_6]^{3+}$ has all of its electrons paired. Each of the ligands, as Lewis bases, contributes a pair of electrons to form a coordinate covalent bond. The valence bond theory designations of the electronic structures are shown in Figure 2–7. The bonding is described as being covalent. Appropriate combinations of metal atomic orbitals are blended together to give a new set of orbitals, called *hybrid orbitals*, that form the most stable covalent bonds between the metal and the ligands.

In six-coordinated systems the hybrid orbitals involve the s, p_x, p_y, p_z, $d_{x^2-y^2}$, and d_{z^2} atomic orbitals. The resulting six sp^3d

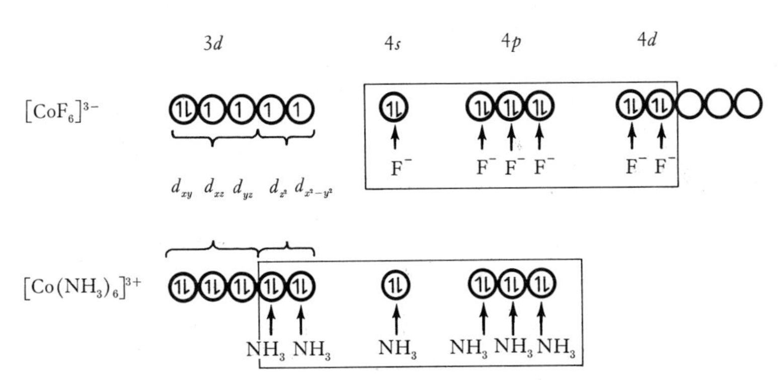

Figure 2–7 Valence bond theory representations of $[CoF_6]^{3-}$ and $[Co(NH_3)_6]^{3+}$.

hybrid orbitals point toward the corners of an octahedron. Note that for $[CoF_6]^{3-}$, the *d* orbitals used are of the same principal energy level as the *s* and *p* orbitals. A complex of the $nsnp^3nd^2$ type is called an *outer-orbital* complex because it uses "outer" *d* orbitals. On the other hand, $[Co(NH_3)_6]^{3+}$ uses *d* orbitals of a lower principal energy level than the *s* and *p* orbitals. Such a complex, $(n-1)d^2ns\ np^3$, is called an *inner-orbital* complex because it uses "inner" *d* orbitals. See Section 2–5 for the nomenclature used in these systems on the basis of the crystal field theory (16).

2–5 ELECTROSTATIC CRYSTAL FIELD THEORY

The valence bond theory and the electrostatic theory are very different. The former starts with the premise that the coordinate bond is covalent. The electrostatic theory completely neglects covalent bonding and assumes that the bond between metal and ligand is totally ionic. Calculations of coordinate bond energies can be made by using classical potential-energy equations that take into account the attractive and repulsive interactions between charged particles (15). In (15), q_1 and q_2 are the charges on the interacting

$$\text{bond energy} = \frac{q_1 q_2}{r} \qquad (15)$$

ions and r is the distance that separates the ion centers. A similar equation applies to the interaction between an uncharged polar molecule and an ion. This approach gives results that are in reasonably good agreement with the experimental bond energies for non-transition-metal complexes. For transition-metal complexes, the calculated values are often too small. This discrepancy is largely corrected when the *d* orbital electrons are considered and allowance is made for the effect of the ligands on the relative energies of the *d* orbitals.

This refinement of the electrostatic theory was first recognized and used by the physicists Bethe and Van Vleck in 1930 to explain the colors and magnetic properties of crystalline solids. Their theory is known as the *crystal field theory* (CFT). Although this theory was proposed at about the same time as—or even a little earlier than—

the VBT, it took about twenty years for the CFT to be recognized and used by chemists. Perhaps the reasons were that the CFT was written for physicists and the VBT gave such a satisfying pictorial representation of the bonded atoms.

In 1951 several theoretical chemists working independently used the CFT to interpret spectra of transition-metal complexes. Since this approach was most successful, there followed an immediate avalanche of research activity in the area. It soon became apparent that CFT is able to explain in a semiquantitative fashion many of the known properties of coordination compounds.

If one is to understand the CFT, it is necessary to have a clear mental picture of the spatial orientation of the *d* orbitals (Figure 2–6). It is the interaction of the *d* orbitals of a transition metal with ligands surrounding the metal that produces CF effects. We can illustrate the CFT by considering the octahedral complex $[TiF_6]^{2-}$. In a free Ti^{4+} ion, one isolated from all other species, the electronic configuration is $1s^2 2s^2 2p^6 3s^2 3p^6$; no *d* electrons are present. The five empty 3*d* orbitals of this ion have identical energies. This means that an electron may be placed in any one of these *d* orbitals with equal ease. Orbitals that have the same energy are called *degenerate orbitals*.

In $[TiF_6]^{2-}$ the Ti^{4+} ion is surrounded by six F^- ions. The presence of these F^- ions will make it much more difficult to place electrons in the Ti^{4+} *d* orbitals; this is due to repulsion of the electrons by the negative charge on the F^- ions. In other words, the energy of the *d* orbitals increases as F^- ions (or other ligands) approach the orbitals (Figure 2–8). If the six F^- ions surrounding Ti^{4+} in $[TiF_6]^{2-}$ were situated equally near each of the five *d* orbitals of Ti^{4+}, all of these *d* orbitals would have the same energy (they would be degenerate), but an energy considerably greater than that which they had in the free Ti^{4+} ion. However, an octahedral complex in which all *d* orbitals remain degenerate is but a hypothetical situation.

The complex $[TiF_6]^{2-}$ has an octahedral structure; for convenience, we shall visualize this complex with the six F^- ions residing on the x, y, and z axes of a cartesian coordinate system (II). In this orientation the F^- ions are very near the $d_{x^2-y^2}$ and d_{z^2} orbitals, which are referred to as the e_g orbitals (Figure 2–6). In fact the e_g orbitals

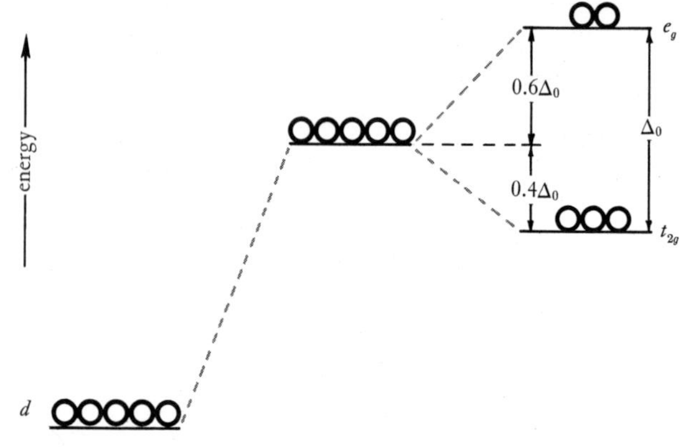

Figure 2–8 The energies of the *d* orbitals in a free metal ion, in a hypothetical complex in which there is no crystal field splitting, and in an octahedral complex.

point directly at the F^- ligands, whereas the d_{xy}, d_{xz}, and d_{yz} orbitals —called the t_{2g} orbitals—point between the ligands.[1] Therefore, it is more difficult to place electrons in e_g orbitals than in t_{2g} orbitals,

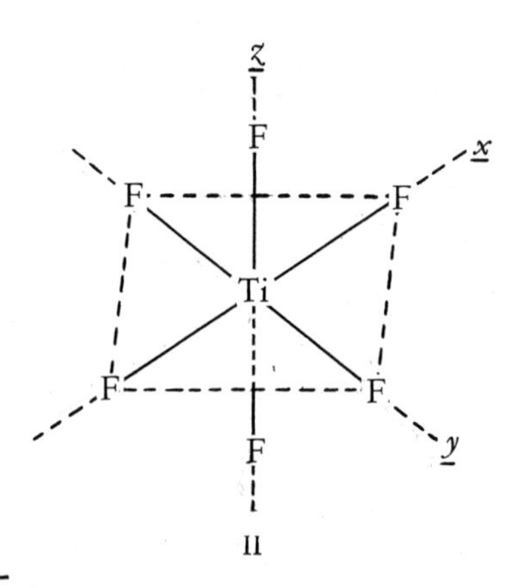

[1] The symbols e_g and t_{2g} are terms used in the mathematical theory of groups. The *t* refers to a triply degenerate set of orbitals; the *e*, a doubly degenerate set.

that is, the e_g orbitals are of higher energy than the t_{2g} orbitals. This conversion of the five degenerate *d* orbitals of the free metal ion into groups of *d* orbitals having different energies is the primary feature of CFT. It is called the *CF splitting*. As we have seen, splitting results because *d* orbitals have a certain orientation in space and because neighboring atoms, ions, or molecules can change the energy of orbitals that are directed toward them in space.

Many students find CFT and its concept of crystal field splitting difficult to visualize. The preceding discussion attempts to describe the essential concepts in simple terms on the basis of the spatial geometries of the *d* orbitals. This is the correct approach to the CFT. It may, however, be helpful to develop a simple physical picture of CF splitting. Refer to Figure 2–9 and imagine that the metal ion with its electron cloud can be represented by a sponge ball. Now consider what happens when a rigid spherical shell (corresponding to ligands) is forced around the outside of the ball. The volume of the ball decreases, and the system has a higher energy, as is evident from the fact that the sponge will expand spontaneously to its original volume upon removal of the constricting shell. This change in energy corresponds to the increase in energy that results from the repulsion between electrons in a metal ion and the electrons of ligands in the hypothetical complex (Figure 2–8).

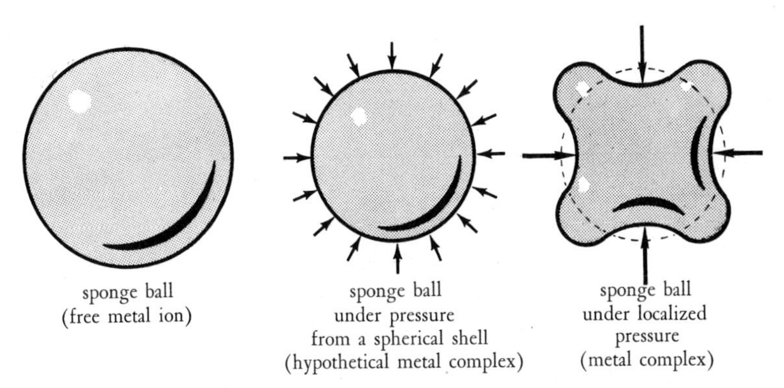

Figure 2–9 Crystal field effects visualized as a sponge ball under spherical pressure and under localized pressure. Compare with Figure 2–8.

Now if the rigid shell is instead allowed to concentrate its total force on six particular spots (for example, the corners of an octahedron), then the sponge is pressed inward at these positions but bulges outward between them. Compared with the spherically constricted system, the sponge at the six points of high pressure is at a higher energy and the sponge at the bulges between is at a lower energy. This corresponds to crystal field splitting with the bulges related to t_{2g} orbitals and the points of depression related to e_g orbitals.

In the preceding discussion it was noted that the energy of the *d* orbitals of a metal ion increases when ligands approach the ion. This in itself suggests that a complex should be less stable than a free metal ion. However, the fact that complexes do form indicates that the complex is a lower energy configuration than the separated metal ion and ligands. The increase in energy of the *d* orbitals of the metal ion is more than compensated for by the bonding between metal ion and ligand.

In an octahedral arrangement of ligands the t_{2g} and e_g sets of *d* orbitals have different energies. The energy separation between them is given the symbol Δ_o. It can be proved in terms of the geometry of an octahedral system that the energy of the t_{2g} orbitals is $0.4\Delta_o$ less than that of the five hypothetical degenerate *d* orbitals that result if CF splitting is neglected (Figure 2–8). Therefore, the e_g orbitals are $0.6\Delta_o$ higher in energy than the hypothetical degenerate orbitals.

In an octahedral complex that contains one *d* electron (for example, $[Ti(H_2O)_6]^{3+}$) that electron will reside in the *d* orbital of lowest energy. Simple electrostatic theory does not recognize that *d* orbitals in a complex have different energies. Therefore, the theory predicts that the *d* electron would have the energy of the hypothetical degenerate *d* orbitals. In fact the *d* electron goes into a t_{2g} orbital that has an energy $0.4\Delta_o$ less than that of the hypothetical degenerate orbitals. Thus the complex will be $0.4\Delta_o$ more stable than the simple electrostatic model predicts. In simple terms we can say that the *d* electron, and hence the whole complex, has a lower energy as a result of the placement of the electron in a t_{2g} *d* orbital that is as far from the ligands as possible. The $0.4\Delta_o$ is called the *crystal field stabilization energy* (CFSE) for the complex.

Table 2–1 gives the crystal field stabilization energies for metal ions in octahedral complexes. Note that the CF stabilization values in Table 2–1 are readily calculated by assigning a value of $0.4\Delta_o$ for each electron put in a t_{2g} orbital and a value of $-0.6\Delta_o$ for each electron put in an e_g orbital. Thus the CFSE for a d^5 system is either $3(0.4\Delta_o) + 2(-0.6\Delta_o) = 0.0\Delta_o$ or $5(0.4\Delta_o) + 0(-0.6\Delta_o) = 2.0\Delta_o$, depending on the distribution of the five electrons in the t_{2g} and e_g orbitals.

Simple electrostatic theory treats a metal ion as a spherical electron cloud surrounding an atomic nucleus. Crystal field theory provides a better model, since it admits that *d* electrons provide a nonspherical electron cloud in order to avoid positions in which the ligands reside. (They provide a nonspherical electron cloud by residing preferentially in the low-energy orbitals that point between

TABLE 2–1

Crystal Field Stabilization Energies for Metal Ions in Octahedral Complexes

d electrons in metal ions	t_{2g}	e_g	*Stabilization,* Δ_o	t_{2g}	e_g	*Stabilization,* Δ_o
1	(↑)( )( )	( )( )	0.4			
2	(↑)(↑)( )	( )( )	0.8			
3	(↑)(↑)(↑)	( )( )	1.2			
4	(↑)(↑)(↑)	(↑)( )	0.6	(↑↓)(↑)(↑)	( )( )	1.6
5	(↑)(↑)(↑)	(↑)(↑)	0.0	(↑↓)(↑↓)(↑)	( )( )	2.0
6	(↑↓)(↑)(↑)	(↑)(↑)	0.4	(↑↓)(↑↓)(↑↓)	( )( )	2.4
7	(↑↓)(↑↓)(↑)	(↑)(↑)	0.8	(↑↓)(↑↓)(↑↓)	(↑)( )	1.8
8	(↑↓)(↑↓)(↑↓)	(↑)(↑)	1.2			
9	(↑↓)(↑↓)(↑↓)	(↑↓)(↑)	0.6			
10	(↑↓)(↑↓)(↑↓)	(↑↓)(↑↓)	0.0			

ligands.) Therefore, CFT explains why simple electrostatic calculations consistently underestimate the stability of transition-metal complexes and compounds; the simple theory neglects the non-spherical electron distribution and the resulting CFSE.

One of the early objections to a simple electrostatic theory of bonding for metal complexes was that it could not explain the formation of square planar complexes. It was argued that if four negative charges are held to a positive central ion by electrostatic forces alone, then the negative charges must be at the corners of a tetrahedron. Only in such a structure can the negative groups attain maximum separation and so experience a minimum electrostatic repulsion. This is correct if the central ion is spherically symmetric. However, such symmetry is not typical of transition-metal ions, because the electrons will reside in low-energy orbitals that point between ligands and do not have spherical symmetry. We shall see in Section 3–1 that the CFT can account for square planar complexes and, furthermore, that it predicts that certain octahedral complexes will be distorted.

We have considered the crystal field splitting for octahedral complexes; now let us consider complexes of other geometries. It is convenient to start with the CF splitting for an octahedral structure and consider how the splitting changes with a change in geometry (Figure 2–10). In going from a regular octahedron to a square planar structure, the change amounts to the removal of any two trans ligands from the octahedron. Generally, we speak of the *xy* plane as the square plane, which means that trans groups are removed from the *z* axis.

If, instead, the ligands on the *z* axis are moved out so that the metal-ligand distance is only slightly greater than it is for the four ligands in the *xy* plane, the result is a tetragonal structure (Figure 2–10). This permits the ligands in the *xy* plane to approach the central ion more closely. Consequently, the *d* orbitals in the *xy* plane experience a greater repulsion from the ligands than they do in an octahedral structure, and we find an increase in the energy of the $d_{x^2-y^2}$ and d_{xy} orbitals (Figure 2–10). At the same time, the *d* orbitals along the *z* axis or in the *xz* and *yz* plane experience a smaller repulsion from the ligands, which are now some distance removed along the *z* axis. This results in a sizable decrease in energy for the d_{z^2}

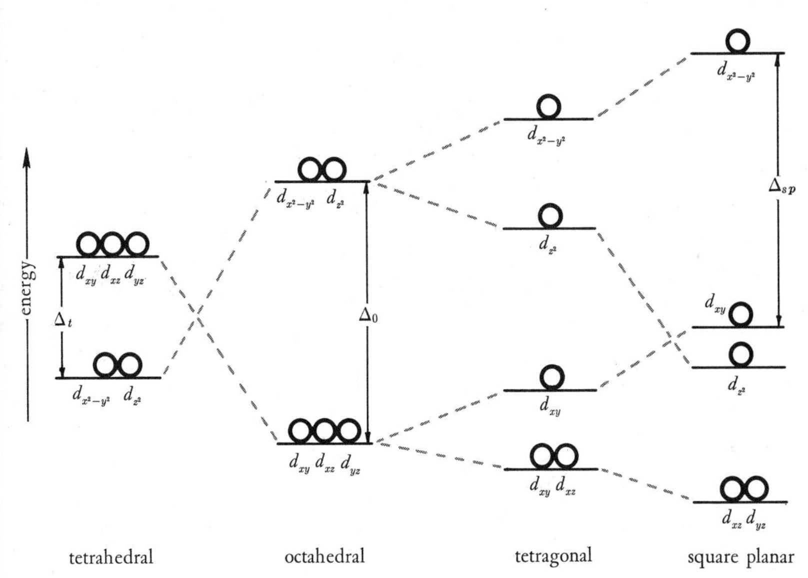

Figure 2–10 Crystal field splittings of the *d* orbitals of a central ion in complexes having different geometries. The subscripts to Δ refer to the geometries.

orbital and a slight decrease for the d_{xz} and d_{yz} orbitals, relative to the octahedral arrangement.

The same splitting pattern is found for a square pyramidal structure, in which there is one ligand on the *z* axis and the other four ligands plus the central atom are in the *xy* plane. The complete removal of the two ligands on the *z* axis to give a square planar configuration is then accompanied by a further increase in energy of the $d_{x^2-y^2}$ and d_{xy} orbitals, as well as a further decrease for the d_{z^2}, d_{xz}, and d_{yz} orbitals.

The CF splitting of the *d* orbitals for a tetrahedral structure is more difficult to visualize. We must first try to picture a tetrahedron placed inside a cube (Figure 2–11). Note that the four corners of the tetrahedron are located at four of the corners of the cube. If we now insert the *x*, *y*, and *z* axes so that they go through the center of the cube and protrude from the centers of its six faces, we can begin to see

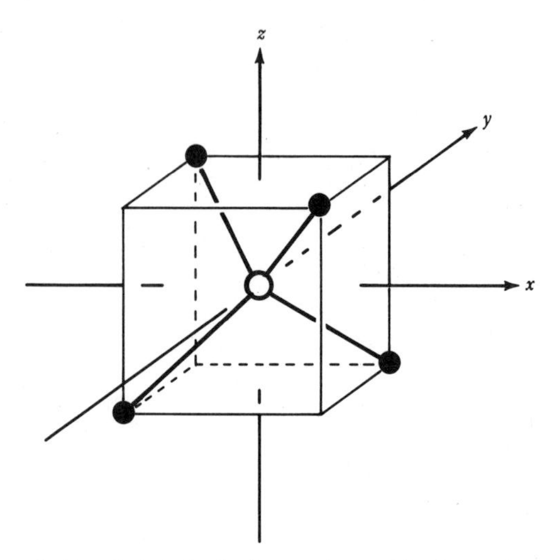

Figure 2–11 A tetrahedral complex with its center at the center of a cube.

the position of the four ligands with respect to the *d* orbitals of the central atom. The *d* orbitals along the cartesian axes ($d_{x^2-y^2}$ and d_{z^2}) are further removed from the four ligands than are the orbitals between the axes (d_{xy}, d_{xz}, and d_{yz}). Therefore, the e_g orbitals ($d_{x^2-y^2}$ and d_{z^2}) are the low-energy *d* orbitals in tetrahedral complexes; the t_{2g} orbitals (d_{xy}, d_{xz}, d_{yz}) are of relatively higher energy. It has been observed that the energy separation between the e_g and t_{2g} orbitals, the crystal field splitting Δ_t, is only about one-half of Δ_o. Hence, crystal field effects favor the formation of octahedral complexes over that of tetrahedral complexes.

The magnetic properties of transition-metal complexes can readily be understood in terms of the crystal field theory. Transition metals have a partially filled *d* sublevel of electrons. If Hund's rule is obeyed, unpaired electrons will be present. For example, a metal ion containing three *d* electrons (called a d^3 system) should have three unpaired electrons (①①①○○); a d^8 metal ion should have two unpaired electrons and three pairs of electrons (⑴⑴⑴①①). Materials that contain unpaired electrons are attracted to a magnet and are said to be *paramagnetic*. (This attraction is much weaker than that exhibited by ferromagnetic materials such as iron.) The magnitude of attraction of a material to a magnet is a measure of the number of unpaired electrons present.

Paramagnetism can be measured with a relatively simple device called a Guoy balance. The sample is placed in a tube suspended from a balance, and the weight of the sample is measured in the presence of a magnetic field and in its absence. If the material is paramagnetic, it will weigh more while the magnetic field is present and attracting it. The increase in weight is a measure of the number of unpaired electrons in the compound.

It has been observed that in some transition-metal complexes Hund's rule is not obeyed. For example, some d^6 cobalt(III) complexes such as $[Co(NH_3)_6]^{3+}$ are not attracted to a magnet (they are *diamagnetic*). Complexes in which some of the unpaired electrons of the gaseous metal ion have been forced to pair are called *low-spin complexes*. The cobalt(III) complex $[CoF_6]^{3-}$ is paramagnetic and contains four unpaired electrons. It is an example of a *high-spin complex*. In such a complex the electron distribution of the complexed metal ion is similar to that found in the gaseous ion. The electron distributions for these two complexes can be represented as (↑↓)(↑↓)(↑↓)( )( ) and (↑↓)(↑)(↑)(↑)(↑), respectively. A variety of names have been given to the behavior for which we have used the terms "high-spin" and "low-spin." These are summarized in (16).

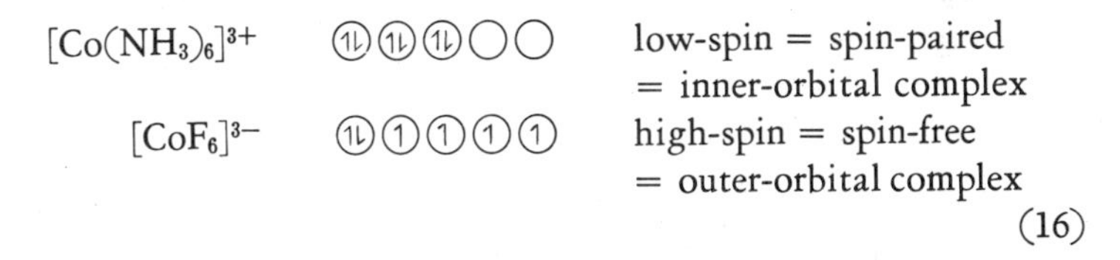

(16)

It is now necessary to try to understand why in such systems the *d* orbital electrons are distributed differently. It must be recognized that at least two effects determine the electron distribution. First, in accordance with Hund's rule the normal tendency is for the electrons to remain unpaired. Energy sufficient to overcome the repulsive interaction of two electrons occupying the same orbital is required to cause the electrons to become paired. Second, in the presence of a CF, the *d* orbital electrons will tend to occupy the low-energy orbitals and thus avoid as much as possible repulsive interaction with the ligands. If the stability thus gained (Δ) is large enough to overcome the loss in stability due to electron pairing, the electrons couple and the result is a *low-spin* complex. Whenever the CF splitting (Δ) is not sufficient, the electrons remain unpaired and the complex is of a *high-spin* type. Note in Figure 2–12 that the value of Δ_o for $[CoF_6]^{3-}$ is smaller than that for $[Co(NH_3)_6]^{3+}$. Complexes in which Δ is large will generally be low-spin complexes. A few additional examples of CF splitting and electron distributions in metal complexes are provided in Figure 2–13.

The magnitude of the CF splitting, as we have seen, determines whether the *d* electrons in a metal ion will pair up or obey Hund's rule. It also influences a variety of other properties of transition metals. The extent of the CF splitting depends on several factors. The nature of the groups (ligands) providing the CF is of greatest interest. From an electrostatic point of view it is clear that ligands with a large negative charge and those that can approach the metal closely (small ions) should provide the greatest CF splitting. Small highly charged ions will make any *d* orbital they approach an energetically unfavorable place to put an electron. This reasoning is in agreement with the observation that the small F^- causes a greater

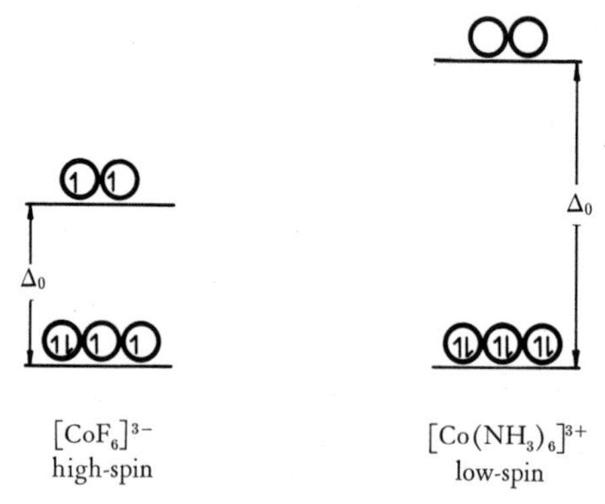

Figure 2–12 Relative crystal field splittings (Δ_o) of the *d* orbitals in high-spin and low-spin octahedral Co(III) complexes.

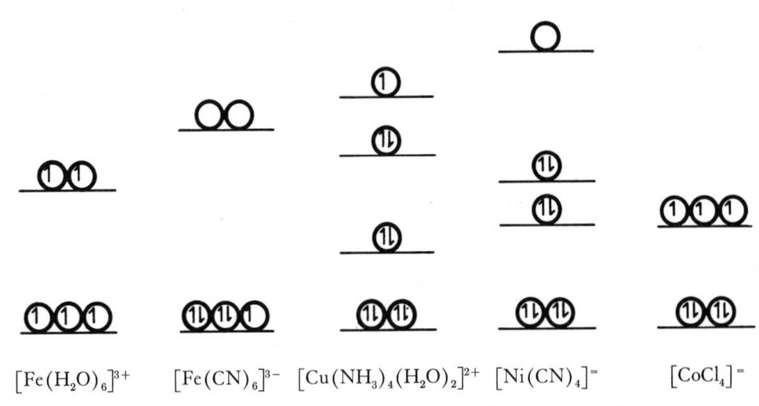

Figure 2-13 Crystal field splittings and electron distributions for some metal complexes. The structures of the first two complexes are octahedral, and the others (left to right) are tetragonal, square planar, and tetrahedral (see Figure 2-10).

crystal field splitting than the larger halide ions Cl^-, Br^-, and I^-.

Since CF splitting arises from a strong interaction of the ligands with orbitals that point directly toward them and a weak interaction with those that point between, in order to achieve a large CF splitting it is desirable that a ligand "focus" its negative charge on an orbital. A ligand with one free electron pair (for example, NH_3) can be visualized as doing this much more readily than a species with two or more free electron pairs, III and IV. This type of argument can be used to account for the observation that neutral NH_3 molecules cause a greater CF splitting than H_2O molecules or negatively charged halide ions.

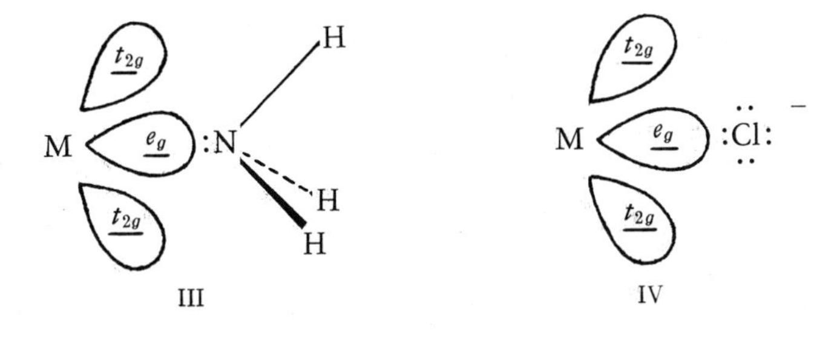

In general, however, it is difficult to explain with a simple electrostatic model the observed ability of various ligands to cause crystal field splitting. The CF splitting ability of ligands has been observed to decrease in the order (17). To account for this order, it is

$$\overbrace{CO, CN^- > phen > NO_2^-}^{\text{strong-field ligands}} > \overbrace{en > NH_3 > NCS^- > H_2O > F^- >}^{\text{intermediate-field ligands}}$$

$$\overbrace{RCO_2^- > OH^- > Cl^- > Br^- > I^-}^{\text{weak-field ligands}} \qquad (17)$$

necessary to abandon a completely ionic electrostatic model for the bonding in complexes and to realize that covalent interactions also exist.

A modified crystal field theory that includes the possibility of covalent bonding is called *ligand field theory*. It can account, at least qualitatively, for the crystal field splitting caused by various ligands. Molecules such as CO, CN^-, phen, and NO_2^-, which provide the largest crystal fields, are all able to form *π bonds* with the central metal atom (Section 2–6). This *π bonding* can markedly increase the magnitude of the CF splitting.

The size of the CF splitting is also strongly influenced by the oxidation state of the metal ion and the type of *d* electrons present. The higher the oxidation state of the metal ion, the larger will be the crystal field splitting. The complex $[Co(NH_3)_6]^{3+}$ is a diamagnetic low-spin complex, whereas $[Co(NH_3)_6]^{2+}$ is a paramagnetic high-spin complex. The crystal field splitting in the Co(III) complex is about twice as great as in the Co(II) complex; this results in the pairing of electrons. One can attribute the larger Δ_o for Co(III) to the fact that the ligands can approach more closely to the smaller, higher-charged metal ion and hence interact more strongly with its *d* orbitals. The CF splitting in $[Rh(NH_3)_6]^{3+}$ and $[Ir(NH_3)_6]^{3+}$ is greater than in $[Co(NH_3)_6]^{3+}$. In general, the crystal field splitting is greatest for complexes containing 5*d* electrons and least for those containing 3*d* electrons. One might attribute this behavior to the fact that 5*d* orbitals extend farther into space and thus interact more strongly with the ligands.

The greatest achievement of CFT is its success in interpreting the colors of transition-metal compounds. One consequence of the comparatively small energy differences Δ between the nonequivalent d orbitals in transition-metal complexes is that the excitation of an electron from a lower to a higher level can be achieved by the absorption of visible light. This causes the complex to appear colored. For example, an aqueous solution of Ti(III) is violet. The color is an indication of the absorption spectrum of the complex $[Ti(H_2O)_6]^{3+}$ (Figure 2–14). That the complex absorbs light in the visible region is explained by the electronic transition of the t_{2g} electron into an e_g orbital (Figure 2–15). The absorption spectra of complexes containing more than one d electron are more complicated because a greater number of electronic transitions are possible.

Planck's equation (18) relates the energy E of an electronic transition to the wavelength λ of the light absorbed. h is Planck's con-

$$E = \frac{hc}{\lambda} \tag{18}$$

stant (6.62×10^{-27} erg/sec) and c is the speed of light (3.00×10^{10} cm/sec). The units of E are ergs per molecule and of λ, centimeters. From Equation (18) it is possible to determine the energy difference

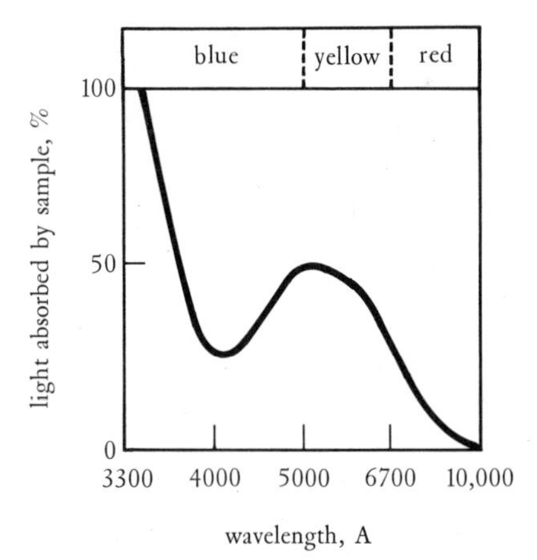

Figure 2–14 The absorption spectrum of $[Ti(H_2O)_6]^{3+}$ Solutions of $[Ti(H_2O)_6]^{3+}$ are red-violet because they absorb yellow light but transmit blue and red.

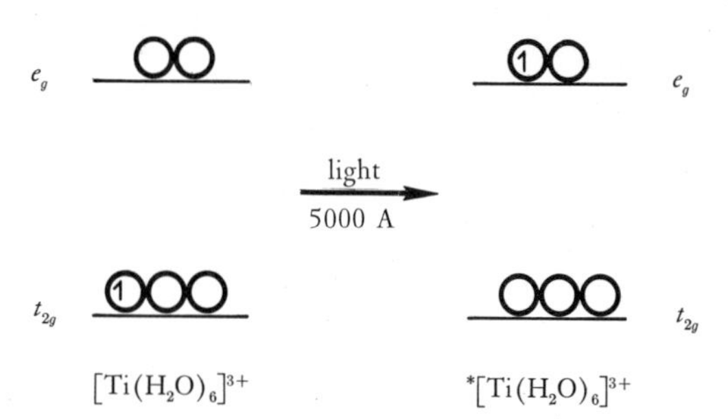

Figure 2-15 The $d - d$ electronic transition responsible for the violet color of $[Ti(H_2O)_6]^{3+}$.

Δ between the d orbitals that are involved in the electronic transition. Collection of the constants h and c plus the use of suitable conversion factors (Avogadro's number, 6.02×10^{23} molecules/mole and Joule's constant, 4.18×10^7 ergs/cal) gives us (19). E has the units kilocalories per mole and λ is in angstroms. The maximum in

$$E = \frac{2.84 \times 10^5}{\lambda} \tag{19}$$

the visible absorption spectrum of $[Ti(H_2O)_6]^{3+}$ is found at a wavelength of 5000 A, giving us a value of 57 kcal/mole for the energy difference between t_{2g} and e_g orbitals. The CF splitting Δ of 57 kcal is of the same order of magnitude as many bond energies. Although this value of 57 kcal is small compared to the heat of hydration of Ti^{3+} (20), 1027 kcal/mole, the CF splitting is very important and

$$Ti^{3+} \text{ (gaseous)} + H_2O \rightarrow [Ti(H_2O)_6]^{3+} \text{ (aqueous)} + 1027 \text{ kcal/mole} \tag{20}$$

necessary to an understanding of transition-metal chemistry.

It should be pointed out that the simple ionic model that is the basis of crystal field theory does not accurately represent the bonding in transition-metal compounds. There is ample experimental evidence that both ionic and covalent bonding play an important role. Nonetheless, the ionic CF theory provides a simple model that will explain a great deal of transition-metal behavior and, moreover, one that has led and will lead to the formulation of many instructive experiments. The role of crystal field theory in the structure, stability, and reactivity of complexes is discussed in later chapters.

2–6 MOLECULAR ORBITAL THEORY

The molecular orbital theory (MOT) is becoming more and more popular with chemists. It includes both the covalent and ionic character of chemical bonds, although it does not specifically mention either. The MOT treats the electron distribution in molecules in very much the same way that modern atomic theory treats the electron distribution in atoms. First the positions of the atomic nuclei are determined. Then orbitals around the nuclei are defined; these molecular orbitals (MO's) locate the region in space in which an electron in a given orbital is most likely to be found. Rather than being localized around a single atom, these MO's extend over part or all of the molecule. Calculations of the shapes of MO's have been made for only the simplest of molecules.

Since calculation of MO's from the first principles is difficult, the usual approach is the *linear combination of atomic orbitals* (LCAO) method. It seems reasonable that the MO's of a molecule should resemble the atomic orbitals (AO's) of the atoms of which the molecule is composed. From the known shapes of AO's, one can approximate the shapes of the MO's. The linear combinations (additions and subtractions) of two atomic *s* orbitals to give two molecular orbitals are pictured in Figure 2–16. One MO results from the addition of the parts of the AO's that overlap, the other from their subtraction.

The MO that results from the addition of the two *s* orbitals includes the region in space between the two nuclei; it is called a *bonding MO*, and it is of lower energy than either of the two *s* AO's from which it arose. The MO that results from subtraction of the parts of the AO's that overlap does not include the region in space between the nuclei. It has a greater energy than the original AO's, and it is called an *antibonding MO*. One can appreciate the energy difference between the bonding and antibonding MO's if one realizes that electrons, when they reside in a region between two nuclei, are favorably influenced by both nuclei. Electrons in antibonding MO's are under the influence of only one nucleus.

Combinations of *s* atomic orbitals give σ (*sigma*) *MO's*. A combination of *p* AO's, as shown in Figure 2–16, may give either σ or

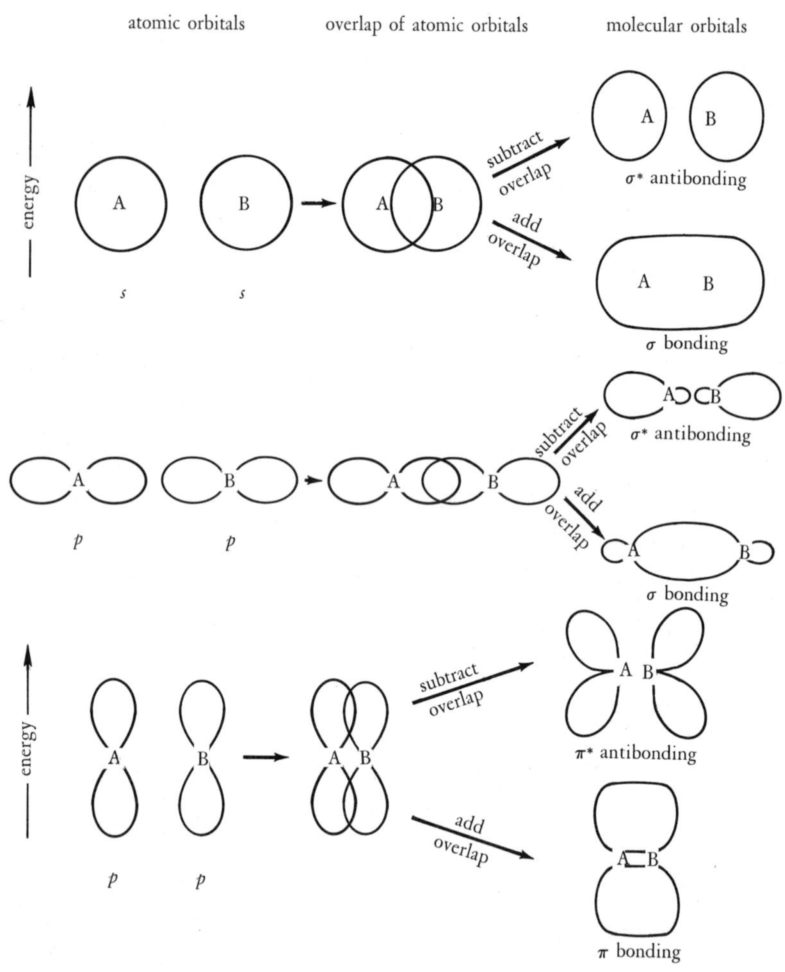

Figure 2–16 The formation of MO's by the LCAO method.

π *(pi) MO's*. In a π MO there is a plane passing through both nuclei along which the probability of finding an electron is zero. Electrons in π MO's reside only above and below the bond axis.

To illustrate the use of MOT, let us look at the MO energy diagrams for a few simple molecules. The H_2 molecule diagram is shown in Figure 2–17. In the separated H atoms one electron resides in each hydrogen AO. In the H_2 molecule both electrons reside in the low-energy σ-bonding MO. *The* H_2 *molecule is more stable than the separated atoms;* the two electrons are both in a lower-energy orbital in the molecule. The *difference between* the *energy of* the *AO's* and the *bonding MO depends* on how much the *AO's overlap* in the molecule. *A large overlap* results in a large difference and hence a *strong bond;* a small overlap results in a small difference, and hence the molecule will be of only slightly lower energy than the separated atoms.

The dihelium ion He_2^+ is a three-electron system; its MO energy level diagram is shown in Figure 2–18. Since an orbital can hold only two electrons, the third electron must go in the σ^* antibonding MO. This orbital is of higher energy than the AO's of the separated He atoms; thus the placing of an electron in the σ^* MO represents a loss of energy and results in a less stable system. This is in agreement with the experimental observation that the He_2^+ bond energy is only 57 kcal/mole compared to 103 kcal/mole for H_2. The four-electron He_2 molecule would be no more stable than two free He atoms.

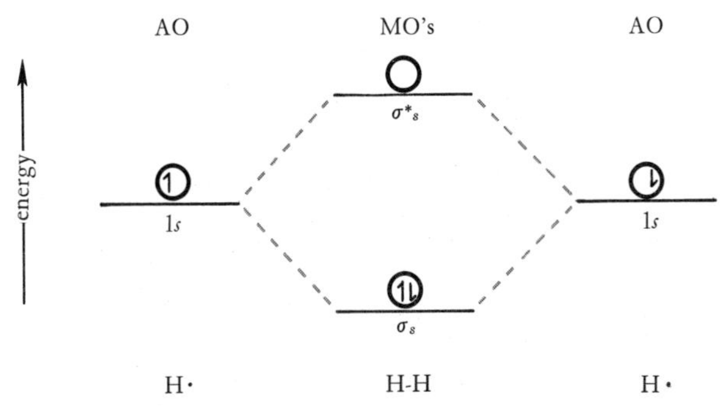

Figure 2–17 The MO diagram for the hydrogen molecule.

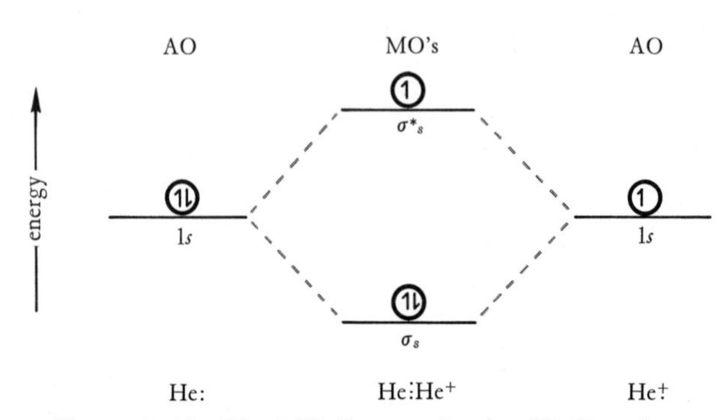

Figure 2–18 The MO diagram for the dihelium ion.

An MO energy level diagram for the general molecule AB is shown in Figure 2–19. There are an infinite number of higher-energy MO's for the AB molecule just as there are an infinite number of higher-energy AO's for the A and B atoms, but the orbitals of interest are the low-energy orbitals in which the electrons reside. When two

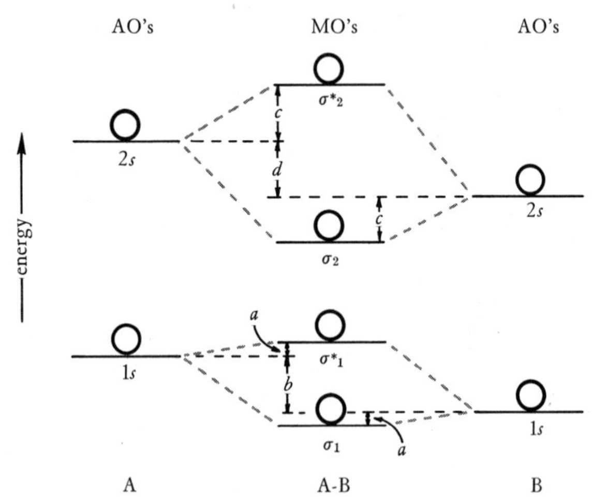

Figure 2–19 The MO diagram for an AB molecule.

different types of atoms are present, the energies of the AO's are expected to differ (for example, the $1s$ orbitals of A and B have different energies). The more electronegative element will have the lowest-energy AO's. The difference in energy between the AO's of the two elements (Figure 2–19*b* and *d*) is a measure of the amount of ionic character in the bond. In H_2 the $1s$ orbitals of the two H atoms have the same energy; hence there is no net ionic character in the bond.

The larger the difference in energy between the two AO's that combine to give an MO, the more ionic is the bond. In the AB molecule the σ_1 MO has an energy near that of the B $1s$ AO; this implies that it resembles the B $1s$ AO more than the A $1s$ AO. If A and B contribute one electron each to the σ_1 MO, this corresponds to a transfer of electronic charge from A to B, since σ_1 resembles B more than A. The magnitudes of a and c are another feature of interest; they depend upon the amount of overlap between the atomic orbitals of A and B and are a measure of the amount of covalent bonding. In Figure 2–19, $a < c$; this implies that the $1s$ orbitals on A and B do not penetrate far enough into space to overlap very much with each other, whereas the $2s$ orbitals can interact more favorably, since they extend farther from the nuclei. The amount of energy released in the formation of an A—B bond depends on the number and energy of the electrons that A and B contribute to the molecule. Table 2–2 illustrates this point.

The MO energy level diagrams for metal complexes are much more complicated than those for simple diatomic molecules. However, in the MO diagrams for $[Co(NH_3)_6]^{3+}$ and $[CoF_6]^{3-}$ in Figure 2–20 one can recognize several familiar features. On the left are $3d$, $4s$, and $4p$ atomic orbitals of Co^{3+}. The lower- and higher-energy AO's are of less interest. Since six ligands are involved, the right side of the diagram is somewhat different from the diagrams we have seen previously. Only one energy level, that of the ligand orbitals used in σ bonding, is shown. (More complicated diagrams are sometimes used.) Since all six ligands are alike, this energy level represents the energy of an orbital from each of six ligands.

The *ligand orbitals* are in general of *lower energy than* the *metal orbitals*, and hence the *bonds have* some *ionic character*. That is, the bonding MO's are more like ligand orbitals than metal orbitals, and

TABLE 2–2

The Amount of Energy Released in the Formation of an AB Molecule[a]

Electrons from A	*Electrons from B*	*Energy released in formation of AB*
$1s^1$	0	$a + b$
0	$1s^1$	a
$1s^1$	$1s^1$	$2a + b$
$1s^2$	0	$2a + 2b$
$1s^2$	$1s^2$	0
$1s^2 2s^1$	$1s^2$	$c + d$
$1s^2$	$1s^2 2s^1$	c
$1s^2$	$1s^2 2s^2$	$2c$
$1s^2 2s^1$	$1s^2 2s^2$	c
$1s^2 2s^2$	$1s^2 2s^2$	0

[a] The reader should supply the data for the situations not included in the table; see Figure 2–19.

placing metal electrons in these MO's thus transfers electronic charge from metal to ligands. Two d orbitals (the e_g orbitals, $d_{x^2-y^2}$ and d_{z^2}), the 4s, and the three 4p orbitals are *oriented* along the x, y, and z axes *where the ligands are located.* Therefore, orbital overlap with the ligand AO's results, and six bonding and six antibonding MO's are formed: σ_s (1), σ_p (3), σ_d (2), σ_d^* (2), σ_s^* (1), σ_p^* (3). The t_{2g} (d_{xy}, d_{xz}, and d_{yz}) orbitals do not point at ligand orbitals and hence are *not involved in σ bonding.* Their energy is unchanged, and they are called nonbonding orbitals.

When the Co(III) and ligand electrons are placed in the complex MO's, we find that the six bonding MO's are filled; this corresponds to six metal-ligand bonds. The remaining electrons are distributed among the nonbonding MO's (the t_{2g} orbitals) and the σ_d^* (antibonding) MO's. The σ_d^* MO's arise from the interaction of metal

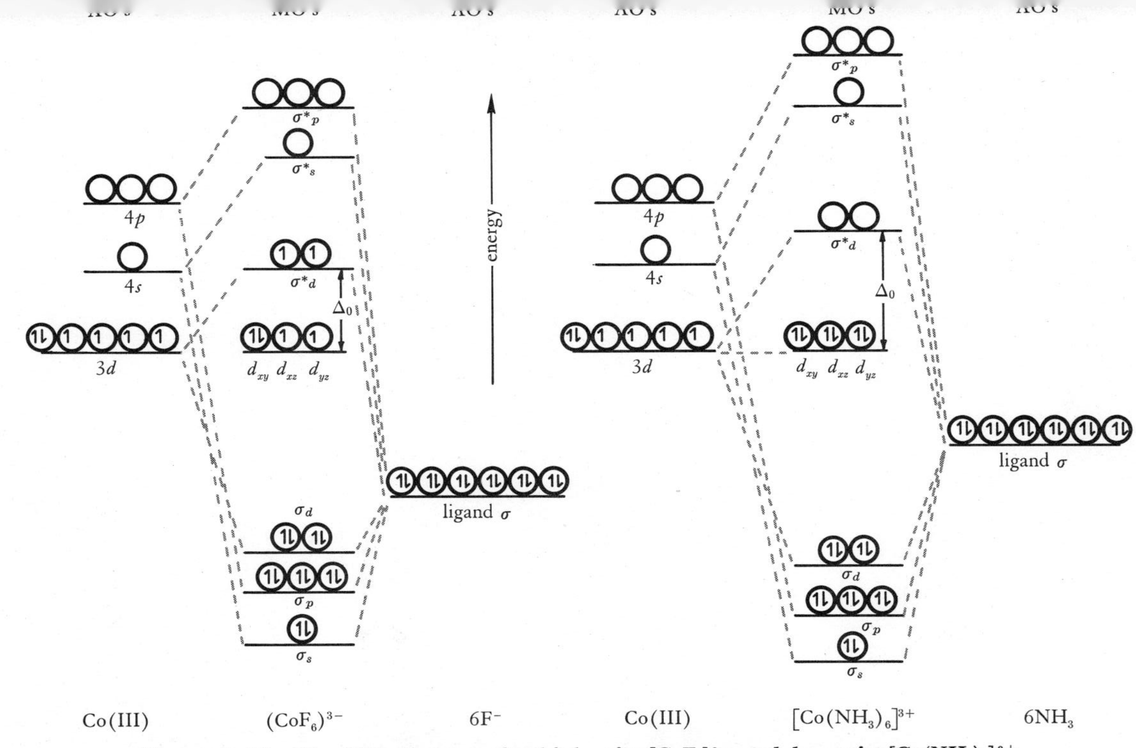

Figure 2–20 The MO diagrams for high-spin $[CoF_6]^{3-}$ and low-spin $[Co(NH_3)_6]^{3+}$.

$d_{x^2-y^2}$ and d_{z^2} orbitals and ligand orbitals, but since the σ_d^* MO's are nearer in energy to the metal $d_{x^2-y^2}$ and d_{z^2} orbitals, they do not differ markedly from them. Therefore the placement of the excess electrons in the t_{2g} and σ_d^* MO's is analogous to the arrangement predicted by the crystal field model, where the same number of electrons is distributed between the t_{2g} and e_g orbitals.

If the difference in energy Δ between the nonbonding t_{2g} orbitals and the σ_d^* MO is small, Hund's rule is obeyed; in $[CoF_6]^{3-}$ this is the case and the d electrons are distributed $t_{2g}{}^4\ \sigma_d{}^{*2}$. The presence of two electrons in the σ_d^* orbitals effectively cancels out the contribution of two electrons in bonding σ_d orbitals and hence weakens the Co—F bonds. When Δ is large as in $[Co(NH_3)_6]^{3+}$, all electrons go into the t_{2g} orbitals. The reasons for the energy separation between the t_{2g} and σ_d^* or e_g orbitals are quite different in the two theories. According to CFT the CF splitting arises from the electrostatic repulsion of d electrons by ligands. MOT essentially attributes the splitting to covalent bonding. The greater the overlap of e_g metal orbitals with the ligand orbitals, the higher in energy will be the σ_d^* orbital.

The MOT can explain the influence of π bonds on the stability of metal complexes and on the magnitude of the CF splitting provided by ligands. Since a quantitative treatment of this subject is quite involved, only a qualitative explanation will be presented here. In the previous discussion it was indicated that the strength of a covalent interaction depends on the extent of overlap of AO's on the two bonded atoms. In previous examples only σ overlap was considered. In $[Fe(CN)_6]^{4-}$ and a variety of other metal complexes both σ and π bonding occur (Figure 2–21). In the σ bond the ligand acts as a Lewis base and shares a pair of electrons with an empty e_g (in Figure 2–21, a $d_{x^2-y^2}$) orbital. In the π bond CN^- ion acts as a Lewis acid and accepts electrons from the filled t_{2g} orbital of the metal (in Figure 2–21, a d_{xy} orbital). The presence of π bonding as well as σ bonding strengthens the metal-ligand bond and contributes to the unusual stability of the $[Fe(CN)_6]^{4-}$ ion. In oxyanions such as MnO_4^-, σ and π bonding are also both important. In this case the ligand (oxygen) provides the electrons for the π bond.

The large CF's that are provided by CN^-, CO, and other π-bonding ligands can be explained in this manner. The t_{2g} orbitals of a metal in an octahedral complex are oriented correctly for π bonding

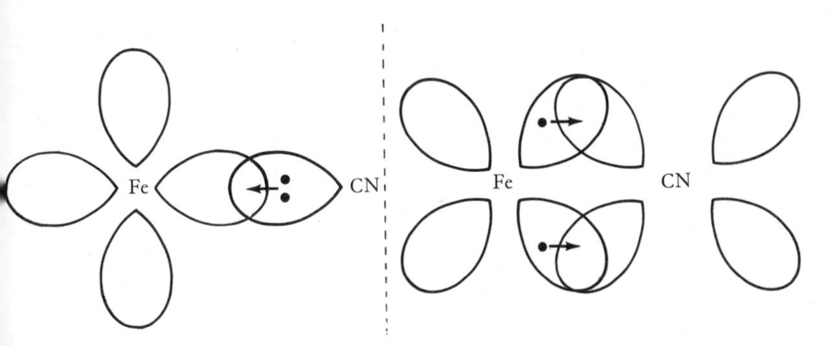

Figure 2–21 **σ and π bonds in $[Fe(CN)_6]^{4-}$. The π bond makes use of a filled *d* orbital of Fe^{2+} and an empty antibonding π^* orbital of CN^- (see π^* in Figure 2–16).**

(Figure 2–21). As was noted previously, the t_{2g} orbitals point between the ligands and hence cannot form σ bonds. In a π bond with a ligand such as CN^-, t_{2g} electrons are partially transferred to the ligand. This process (a bonding interaction) lowers the energy of the t_{2g} orbitals. In Figure 2–8 one can see that a process that will lower the energy of the t_{2g} orbitals must increase Δ_o.

The preceding discussion is a simplified MO approach to bonding, but it illustrates some of the basic ideas and a little of the usefulness of the theory. Molecular orbital theory is very effective in handling both the covalent and ionic contributions to the metal ligand bond.

In conclusion it should be made clear that all three of these theories are at best only good approximations. All three can account qualitatively for many features of metal complexes; all three are used currently, and one or the other may be most convenient for a given application. The most versatile and perhaps most nearly correct is the MOT. Unfortunately, it is also the most complicated and does not lend itself to a pictorial representation of the chemically bonded atoms.

PROBLEMS

1. Determine the EAN of the metal in each of the following compounds. Note that several of these metals do not have an EAN equal to the atomic number of a rare gas.

$[IrCl_6]^{2-}$, $[Cr(NH_3)_4(H_2O)_2]^{2+}$, $[Ni(NH_3)_6]^{2+}$,
$[Ag(NH_3)_2]^{+}$, $[Cr(CN)_6]^{3-}$, $[Fe(CN)_6]^{4-}$,
$[Co(NH_3)_2(NO_2)_4]^{-}$, $[Cr(CO)_6]$, $[Tc_2(CO)_{10}]$,
$[Fe(CO)_2(NO)_2]$, $[Be(H_2O)_4]^{2+}$, $[Al(C_2O_4)_3]^{3-}$

2. The crystal field splittings of *d* orbitals that arise from tetrahedral, tetragonal, and octahedral arrangements of ligands have been described. Predict the splitting that will be produced for the following complexes and structures:

MX_2 linear complex
MX_3 planar complex $\angle XMX = 120°$
MX_6 four long M—X bonds, two short MX bonds in an octahedral transconfiguration
MX_8 a square prism structure

3. What is the CFSE for the following systems?

d^1 octahedral, d^5 low-spin octahedral,
d^8 high-spin octahedral, d^1 tetrahedral,
d^5 high-spin tetrahedral

4. Represent the electronic configuration of the following complexes by using each of the three bonding theories, VBT, CFT, and MOT: $[Fe(H_2O)_6]^{2+}$ (high-spin), $[Ni(NH_3)_6]^{2+}$ (high-spin), and $[Co(C_2O_4)_3]^{3-}$ (low-spin).

REFERENCES

L. Pauling, *The Nature of the Chemical Bond*, 3d ed., Cornell, Ithaca, N.Y., 1960. A complete discussion of VBT is found in this book.

L. E. Orgel, *An Introduction to Transition Metal Chemistry: Ligand-Field Theory*, Wiley-Interscience, New York, 1960.

M. C. Day, Jr., and J. Selbin, *Theoretical Inorganic Chemistry*, Reinhold, New York, 1962.

C. J. Ballhausen, *Introduction to Ligand Field Theory*, McGraw-Hill, New York, 1962.

R. G. Pearson, "Crystal field explains inorganic behavior," *Chem. Eng. News*, **37,** No. 26, 72 (1959).

L. E. Sutton, "Some recent developments in the theory of bonding in complex compounds of the transition metals," *J. Chem. Educ.*, **37,** 498 (1960).

A. D. Liehr, "Molecular orbital, valence bond, and ligand field," *J. Chem. Educ.*, **39,** 135 (1962).

L. Pauling, "Valence bond theory in coordination chemistry," *J. Chem. Educ.*, **39,** 461 (1962).

H. B. Gray, "Molecular orbital theory for transition metal complexes," *J. Chem. Educ.*, **41,** 2 (1964).

III

Stereochemistry

Stereochemistry is that branch of chemistry concerned with the structures of compounds. The liberal use of simple stick models (any do-it-yourself type is fine) is highly recommended as a visual aid to the study of three-dimensional structures. Stereochemistry is sometimes considered to be solely a province of organic chemistry, but this is a gross misapprehension. Because of the unique extent to which carbon forms carbon-carbon chains, organic compounds have a large variety of shapes and structures. However, if attention is focused on an individual carbon atom, the four groups surrounding it are located at the corners of a tetrahedron. Furthermore, since carbon is a second-row element, only the *s* and *p* orbitals are available for bond formation.

Inorganic stereochemistry deals with central atoms having coordination numbers from two to nine. In inorganic compounds it is often necessary to consider not only *s* and *p* orbitals but also *d* and even *f* orbitals. Isomerism, sometimes similar to that of organic compounds and sometimes different, is common in metal complexes.

3–1 GEOMETRY OF COORDINATION COMPOUNDS

Metal complexes are observed to have a variety of structures. Silver complexes are often linear; beryllium complexes are usually tetrahedral; iron forms carbonyl compounds that have trigonal bipyramid structures; cobalt(III) complexes are invariably octahedral; and tantalum forms an eight-coordinated fluoride complex (Figure 3-1). Although a variety of coordination numbers and structures have been observed in metal complexes, the only common coordination numbers are four and six; the common structures corresponding to these coordination numbers are tetrahedral and square planar, and octahedral, respectively. In a study of metal complexes it soon becomes clear that the octahedral structure is by far the most common of these configurations.

An interesting and useful approach to the prediction of structure of compounds in which the coordination number of the central atom is known is given by the VSEPR (valence shell electron-pair repul-

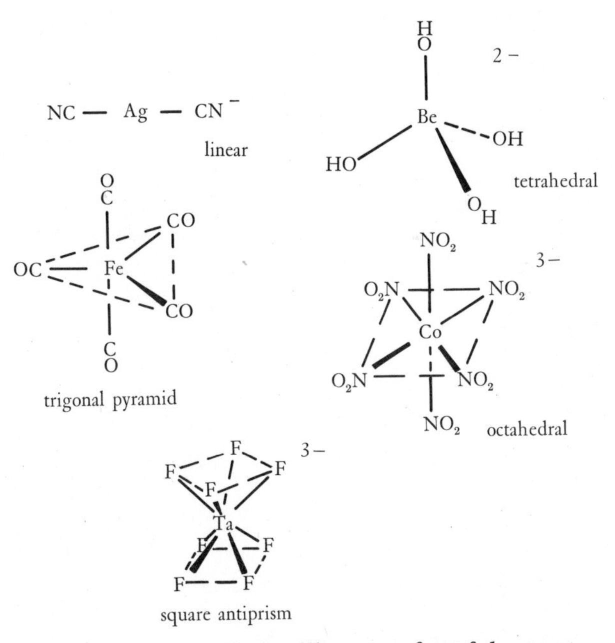

Figure 3-1 Compounds that illustrate a few of the structures found in metal complexes.

sion) theory of Gillespie and Nyholm.[1] One conclusion from this theory is that in general, four- and six-coordinated complexes will be tetrahedral and octahedral, respectively. Complexes of transition metals sometimes deviate from this rule, and the deviation can be attributed to the presence of *d* electrons. Crystal field theory provides perhaps the simplest explanation of the effect of *d* electrons on the structure of complexes.

The CFT claims that *d* orbitals have a specific geometry and orientation in space and that *d* electrons will reside in the orbitals that are farthest from neighboring atoms or molecules. The presence of *d* electrons in six- or four-coordinated complexes tends to cause distortion of the expected octahedral or tetrahedral configuration. *The distortion arises because ligands will avoid those areas around a metal ion in which the d electrons reside.* For example, in $[Ti(H_2O)_6]^{3+}$ there are six water molecules around Ti^{3+}; hence, an octahedral distribution of ligands is expected.

Next one must consider the influence of the metal *d* electrons on the structure. If there were zero, five (unpaired), or ten *d* electrons present in the outer *d* sublevel, the *d* electrons would cause no distortion. A filled sublevel of ten *d* electrons has spherical electrical symmetry; a charged particle (for example, a ligand) on a sphere having the metal at its center will encounter the same electrostatic force regardless of its position on the sphere. If there is one electron in each of the five *d* orbitals, the metal ion also has spherical symmetry. Therefore, the position a ligand will occupy is not influenced by *d* electrons in these cases.

The complex $[Ti(H_2O)_6]^{3+}$ contains one *d* electron; this electron will repel ligands that are near it. From CFT we know that the one electron will be in a low-energy t_{2g} orbital that points between the H_2O ligands. If the electron is assigned to a d_{xy} orbital, one would expect the predicted octahedral structure to distort. Since the d_{xy} orbital lies closest to the four ligands in the *xy* plane, one would predict that these ligands would move away from the metal ion; the structure would become tetragonal with two groups closer to the metal ion than the other four. The same result would be expected if

---

[1] R. J. Gillespie, "VSEPR theory," *J. Chem. Educ.*, **40**, 295 (1963).

the electron were placed in a d_{xz} or d_{yz} orbital. (The reader should convince himself of this.)

Since the t_{2g} orbitals point between ligands, one might expect that the effect of the presence of an electron in one of these orbitals would be small. In fact there is no experimental evidence for tetragonal distortion in $[Ti(H_2O)_6]^{3+}$ or other d^1 systems. In octahedral complexes containing two or three *d* electrons these electrons reside in t_{2g} orbitals that point between ligands. Although one would expect a small distortion for octahedral d^2 systems, there is again no experimental evidence for it. In octahedral d^3 complexes such as $[Cr(H_2O)_6]^{3+}$ each t_{2g} orbital contains one electron. From Figure 2–6 one can see that each of the six ligands in an octahedral array would be near two of these *d* electrons, and hence all would experience the same repulsion. No distortion is expected or observed.

In $[Cr(H_2O)_6]^{2+}$, which is a d^4 high-spin system, the first three electrons go in t_{2g} orbitals and produce no distortion of an octahedral structure. The fourth electron goes in an e_g orbital that points directly at ligands. If the electron resides in a d_{z^2} orbital, the ligands on the *z* axis are repelled; if it resides in the $d_{x^2-y^2}$ orbital, the four ligands in the *xy* plane are repelled. In fact, six-coordinated d^4 metal complexes have been found to have distorted structures in all cases studied. For example, in MnF_3 each Mn(III) is surrounded by six F^- ions so arranged that four are closer to the Mn^{3+} ion than the other two are (Figure 3–2).

We have now considered the distortion of octahedral structures

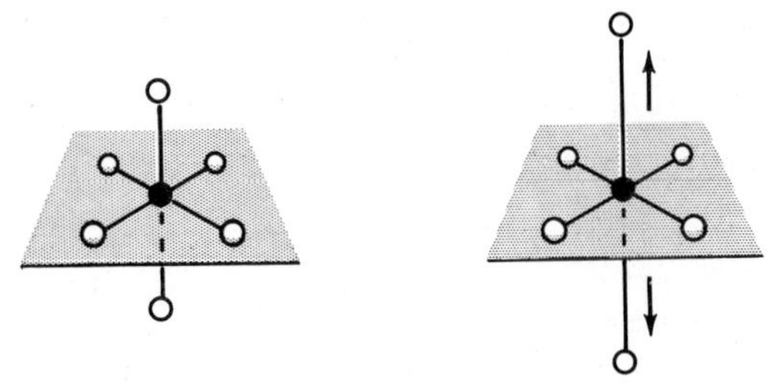

Figure 3–2 An example of a Jahn-Teller distortion.

caused by the presence of 0, 1, 2, 3, 4, 5 (unpaired), and 10*d* electrons. It should be clear that high-spin d^6, d^7, d^8, and d^9 systems are similar to d^1, d^2, d^3, and d^4 systems, respectively. (The first five electrons provide a spherically symmetric electron cloud; the remaining electrons provide the distortion.) Six-coordinated complexes of d^9 metal ions have marked tetragonal distortions similar to those of d^4 complexes. The most common examples are Cu(II) complexes. In $[Cu(NH_3)_4]^{2+}$ the tetragonal distortion is so marked that the square planar tetraammine complex results. It should be noted, however, that solvent molecules occupy the positions above and below the plane in solutions of complexes of this type; these solvent molecules are farther from the metal ion than are the groups in the square plane. The distortion of symmetrical structures resulting from partially filled electronic energy levels (in this case the *d* sublevel) are called *Jahn-Teller* distortions.

The distortions of octahedral structure observed in important low-spin configurations should also be considered. Low-spin d^6 systems are similar to d^3 complexes. The six electrons completely fill the t_{2g} orbitals. Since each of the six ligands is close to two of these orbitals, there is no tendency for distortion, and octahedral structures are observed. Low-spin d^8 complexes are similar to d^4 systems. The last two electrons go into one e_g orbital and interact strongly with the ligands that face this orbital. Marked distortions occur such that two ligands are much farther removed from the central metal than are the other four. In fact, low-spin d^8 complexes are almost invariably square planar. The distortions that result from the presence of *d* electrons in "octahedral" complexes are summarized in Table 3–1.

We have considered the distortions to octahedral structure that result from the presence of *d* electrons. Tetrahedral structures are also observed in metal complexes; however, they are less common than octahedral and distorted octahedral configurations. If four ligands surround a metal atom, a tetrahedral structure is expected. The presence of *d* electrons may then result in distortion of the tetrahedron.

Two exceptions must be noted. As we have seen, four-coordinated low-spin d^8 complexes are square planar, as are four-coordinated d^9 and high-spin d^4 complexes. Metal complexes containing

TABLE 3-1

The Distortions of Octahedral Structures that Result from the Presence of d Electrons

System	*Predicted structure*	*Comments*
High spin:		
d^1, d^6	Tetragonal distortion	Not observed
d^2, d^7	Tetragonal distortion	Not observed
d^3, d^8	No distortion	Experimentally verified
d^4, d^9	Large tetragonal distortion	Experimentally verified
d^5, d^{10}	No distortion	Experimentally verified
Low-spin:		
d^6	No distortion	Experimentally verified
d^8	Large tetragonal distortion	Square planar compounds

0, 5 unpaired, and 10 d electrons are undistorted, as was noted previously. As in octahedral complexes the placement of electrons in orbitals pointing between ligands provides no observable distortion; thus tetrahedral d^1, d^2, d^6, and d^7 complexes appear to be undistorted. The remaining tetrahedral systems, d^3, d^4, d^8, and d^9, should exhibit marked Jahn-Teller distortions. Very few examples of compounds of this type exist, however. Low-spin tetrahedral complexes need not be discussed, since there are no examples of such complexes. The tetrahedral CF splitting (Δ_t) is apparently too small to cause spin pairing.

Although it is possible to predict fairly accurately the stereochemistry of complex ions in which the coordination number of the central atom is known, it is much more difficult to predict the coordination number of the central atom. Large coordination numbers are favored by the electrostatic attraction of negatively charged ligands (or polar molecules) for a positive metal ion. Covalent bonding theories predict in general that the greater the number of bonds formed to an element the greater is the stability of the resulting compound.

The tendency for large coordination numbers is opposed by steric and electrostatic (or Pauli) repulsion between ligands. No simple scheme has been presented to make predictions from these criteria. It might be noted, however, that the first-row transition elements are frequently six-coordinated. Four-coordination is observed primarily in complexes containing several large anions, such as Cl^-, Br^-, I^-, and $O^=$, or bulky neutral molecules. The second- and third-row transition elements exhibit coordination numbers as large as eight.

3–2 ISOMERISM IN METAL COMPLEXES

Molecules or ions having the same chemical composition but different structures are called *isomers*. The difference in structure is usually maintained in solution. Isomers are, therefore, not merely different crystalline forms of the same substance; for example, the rhombic and monoclinic forms of sulfur are not isomers. Metal complexes exhibit several different types of isomerism; the two most important are geometrical and optical. Other types also will be described, and specific examples for each will be given. One fact to note is that, in general, only complexes which react slowly are found to exhibit isomerism. This is because complexes that react rapidly often rearrange to yield only the most stable isomer (Chapter VI).

3–3 GEOMETRICAL ISOMERISM

In metal complexes the ligands may occupy different types of positions around the central atom. Since the ligands in question are usually either next to one another (*cis*) or opposite each other (*trans*), this type of isomerism is often also referred to as *cis-trans isomerism*. Such isomerism is not possible for complexes of CN 2 or 3 or for tetrahedral complexes. In those systems all coordination positions are adjacent to one another. However, cis-trans isomerism is very common for square planar and octahedral complexes, the only two types to be discussed here. Methods of preparation and reactions of some of these compounds are described in Chapter IV.

Platinum(II) complexes are very stable and slow to react; among them are numerous examples of square planar geometrical isomers. No doubt the best known of these are *cis-* and *trans-* $[Pt(NH_3)_2Cl_2]$, I and II. The chemistry of platinum(II) complexes

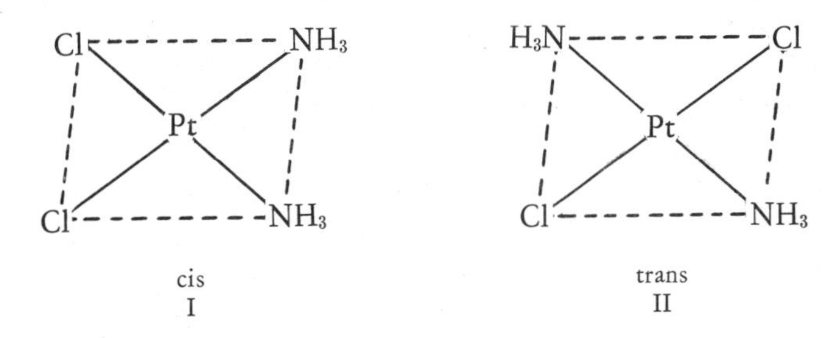

has been studied extensively, particularly by Russian chemists. Many compounds of the types *cis-* and *trans-*$[PtA_2X_2]$, $[PtABX_2]$, and $[PtA_2XY]$ are known. (A and B are neutral ligands such as NH_3, py, $P(CH_3)_3$, and $S(CH_3)_2$; X and Y are anionic ligands such as Cl^-, Br^-, I^-, NO_2^-, and SCN^-.) Isomers can be readily distinguished by x-ray diffraction techniques. Other methods of determining the structures of geometrical isomers are discussed in Section 4–9.

A few compounds of platinum(II) containing four different ligands, [PtABCD], are known. Realizing that either B, C, or D groups may be trans to A, it is apparent that there are three isomeric forms for such a compound. The first complex of this type to be obtained in three forms was the cation $[PtNH_3(NH_2OH)pyNO_2]^+$, which has structures III, IV, and V.

In order to designate the structure of a particular isomer, it is

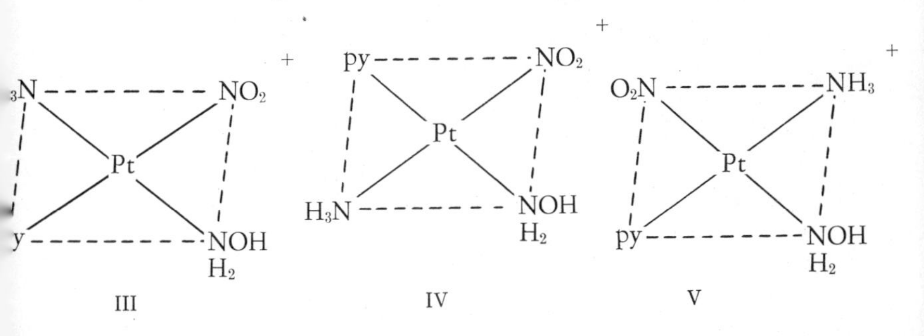

convenient to place the two sets of trans ligands in separate angular brackets ($<>$); for example, $[M < AB > < CD >]$ signifies that A and B are in trans positions, as must be C and D. Individual isomers can be named by the number system (Section 1–3) or by using the trans prefix, which implies that the first two ligands in the name are in trans positions. It follows that the last two ligands named are also in positions trans to each other.

Geometrical isomerism is also found in square planar systems containing unsymmetrical bidentate ligands, $[M(AB)_2]$. Glycinate ion, $NH_2CH_2COO^-$, is such a ligand; it coordinates with platinum(II) to form *cis*- and *trans*-$[Pt(gly)_2]$ having structures VI and VII.

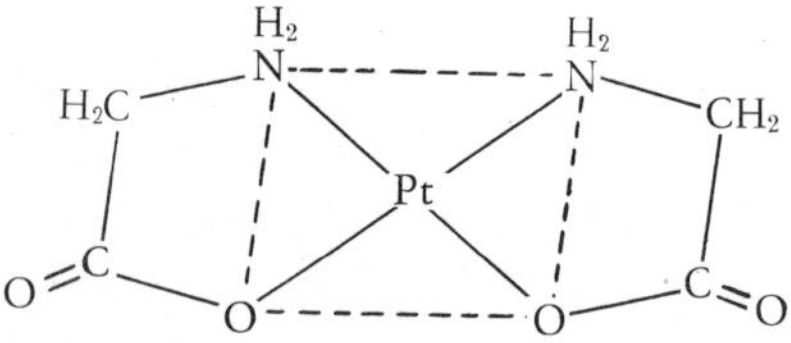

cis-diglycinatoplatinum(II)
VI

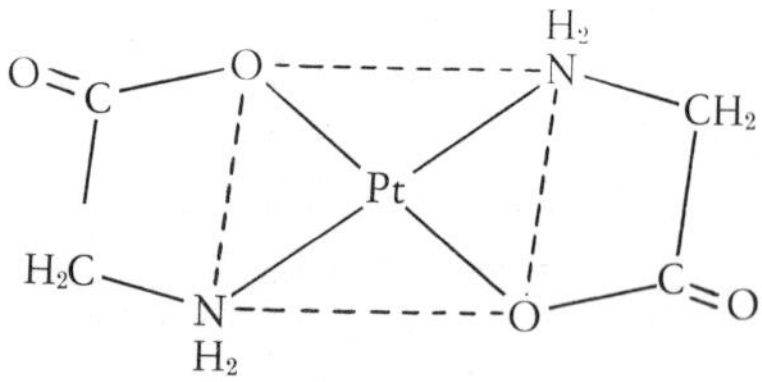

trans-diglycinatoplatinum(II)
VII

It is not necessary that the attached ligand atoms differ; all that is required is that the two halves of the chelate ring be different.

Geometrical isomerism in octahedral compounds is very closely related to that in square planar complexes. Among the most familiar examples of octahedral geometrical isomers are the violet (cis) and green (trans) forms of the dichlorotetraamminecobalt(III) and chromium(III) cations, which have structures VIII and IX. Hundreds of isomeric compounds of the types $[MA_4X_2]$, $[M(AA)_2X_2]$, $(MA_4XY]$, and $[M(AA)_2XY]$, where M = Co(III), Cr(III), Rh(III), Ir(III), Pt(IV), Ru(II), and Os(II), have been prepared and characterized. A few isomers of the type $[MA_3X_3]$ are known; these compounds can form only two geometrical isomers. For example, the isomers of $[Rh(py)_3Cl_3]$ have the structures X and XI. Either the like groups occupy the corners of one of the octahedral faces (cis isomer) or they do not (trans isomer). The largest number of geo-

cis-dichlorotetraammine-
chromium(III) ion
VIII

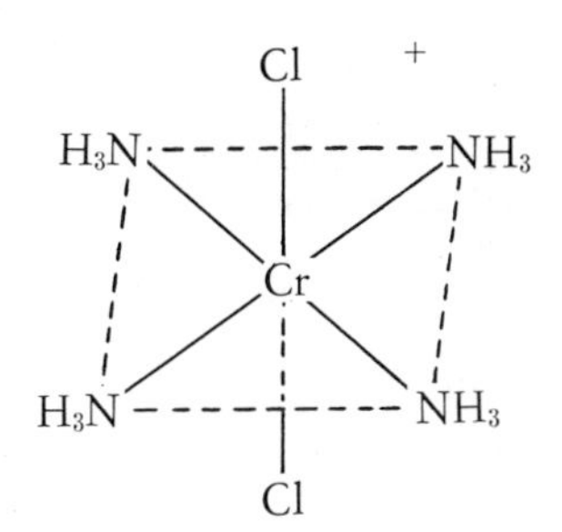

trans-dichlorotetraammine-
chromium(III) ion
IX

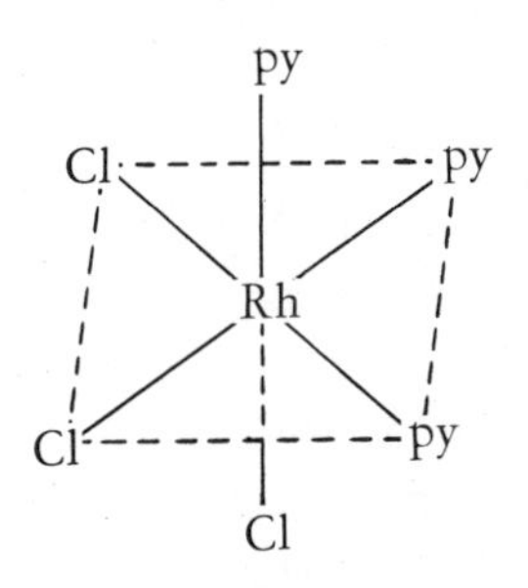

cis-trichlorotripyridinerhodium(III)
X

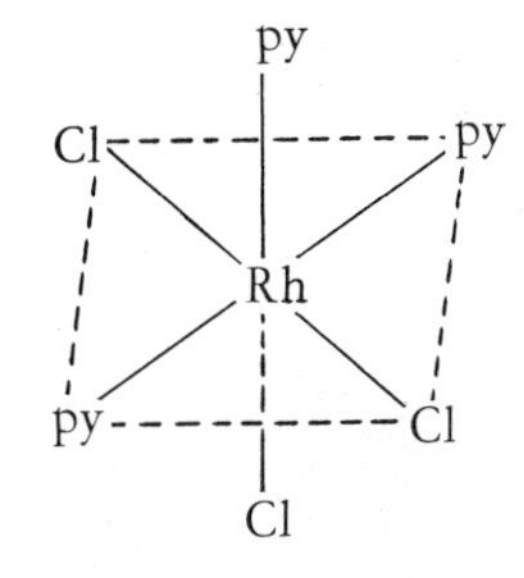

trans-trichlorotripyridinerhodium(III)
XI

metrical isomers would exist for a complex of the type [MABCDEF], wherein each ligand is different. Such a species can exist in 15 different geometrical forms (each form would also have an optical isomer, see Section 3–4). The student may wish to draw all of the possible structures. The only compound of this type that has been prepared is $[Pt(py)(NH_3)(NO_2)(Cl)(Br)(I)]$. It was obtained in three different forms but no attempt was made to isolate all fifteen isomers.

Unsymmetrical bidentate ligands give rise to geometrical isomers in much the same way as was described earlier for square planar complexes. For example, the cis-trans isomers of triglycinatochromium(III) have the structures XII and XIII.[1] Each of these complexes is optically active, as is discussed in the following section.

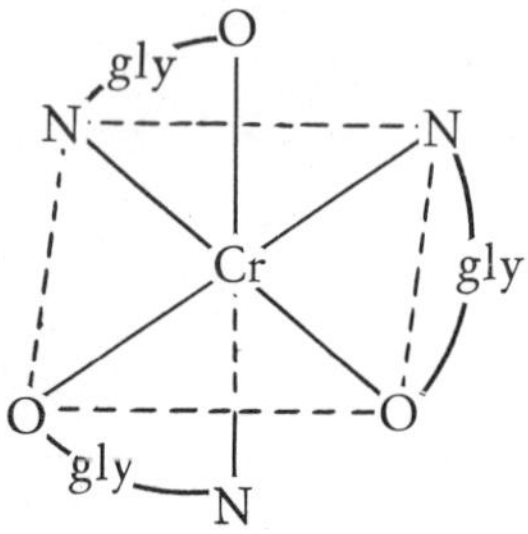

cis
XII

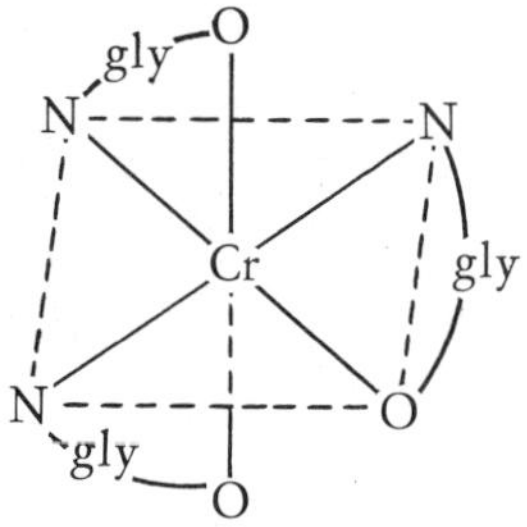

trans
XIII

3–4 OPTICAL ISOMERISM

It has already been necessary to make some references to the phenomenon of optical isomerism. A brief discussion is provided here, along with a few additional examples of optically active metal complexes. Optical isomerism has been recognized for many years. The classical experiments in 1848 of Louis Pasteur, one of the most illustrious and humane of all men of science, showed that sodium ammonium tartrate exists in two different forms. Crystals of the two forms differ, and Pasteur was able to separate them by the laborious task of hand picking.

Aqueous solutions of the two isomers had the property of rotating a plane of polarized light (a beam of light vibrating in only one plane) either to the right or to the left. Because of this property the isomers are said to be *optically active* and are called *optical isomers;* one is designated as the *dextro* (*d*) isomer and the other as the *levo* (*l*) isomer. The extent of rotation of the plane of polarized light by the two isomers is exactly the same; however, the dextro isomer rotates the plane of light to the right, the levo isomer to the left. It follows

---

[1] It is convenient to use abbreviations for chelating ligands in diagrams. In this text the presence of a chelating group is indicated by a curved line on which an abbreviation for the ligand is written. The type of atom bonded to the metal is also indicated.

that the rotations cancel each other in solutions containing equal concentrations of the two isomers. Such a *d, l mixture* is called a *racemic mixture*. Since its solution does not rotate a plane of polarized light, it is optically inactive.

What property of a molecule or ion renders it optically active? The answer is asymmetry (lack of symmetry). The symmetry relationship of optical isomers is similar to that of the right and left hands, or feet, or gloves, or shoes. There is a rather subtle difference between the structures; the relative positions of the thumb and fingers on each hand are the same, yet the two hands are different. One is the mirror image of the other. An analogous situation must exist if a molecule or ion is to be optically active. In order for a molecule or ion to be optically active, it must not have a plane of symmetry; i.e., it should not be possible to divide the particle into two identical halves. Another test that can be applied in attempting to decide whether a given structure will be optically active is to compare it with its mirror image. If the structure and its mirror image are different, the structure will exhibit optical activity.

The *d* and *l* isomers of a given compound are called *enantiomorphs* or *enantiomers*, which mean "opposite forms." In general, they have identical chemical and physical properties. They differ only in the direction in which they rotate a plane of polarized light. This property permits them to be readily detected and to be distinguished. A rather simple instrument known as a *polarimeter* is used for this purpose.

It is interesting to note that sometimes the physiological effects of enantiomers are profoundly different. Thus the *l*-nicotine that occurs naturally in tobacco is much more toxic than the *d*-nicotine that is made in the laboratory. Specific effects such as these are attributed to asymmetric reaction sites in biological systems. Since enantiomers are so similar, and since in chemical reactions the two forms are always produced in equal amounts, special techniques are required to separate the two. This separation process is called *resolution*. Some resolution methods are described in Section 4–10. Often, a single optical isomer will rearrange to give a racemic mixture; the process is called *racemization*.

The simplest possible example of an asymmetric molecule is one with a tetrahedral structure wherein the central atom is surrounded by four different atoms or groups. There are many examples of such molecules among organic compounds. The structures of optical isomers may be represented by the amino acids XIV and XV. Tetrahedral metal complexes are generally very reactive, which makes it extremely difficult to isolate them in isomeric forms. The first example of a tetrahedral metal complex containing four different ligands was reported in 1963, and its resolution has not been achieved. However, complexes containing two unsymmetrical bidentate ligands can be resolved into optically active forms. Optically active isomers of this type are known for complexes of Be(II), B(III), and Zn(II). The enantiomers of bis(benzoylacetonato)beryllium(II) have the structures XVI and XVII. Note that four different groups

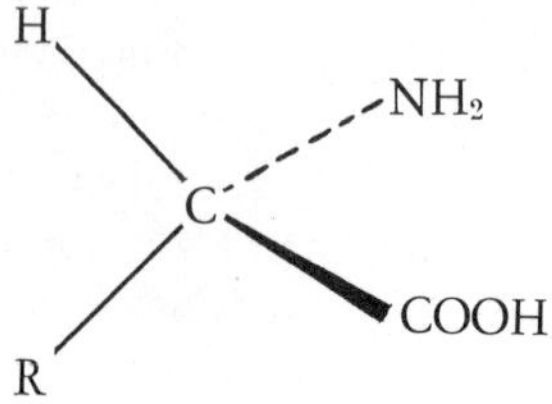

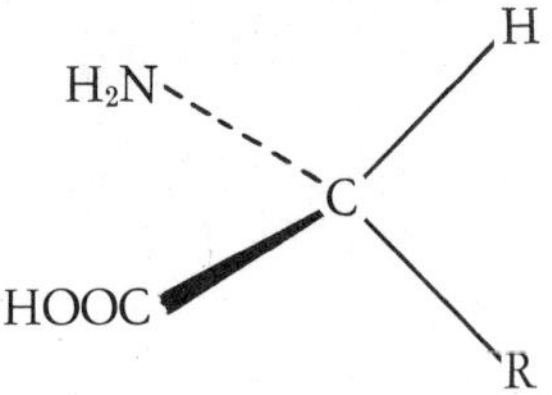

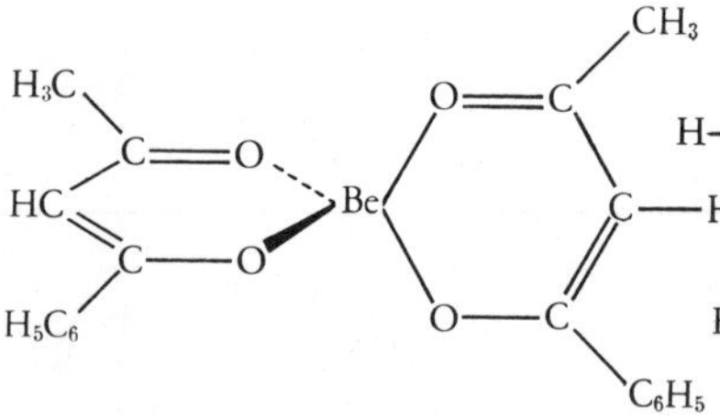

levo
XIV

dextro
XV

XVI

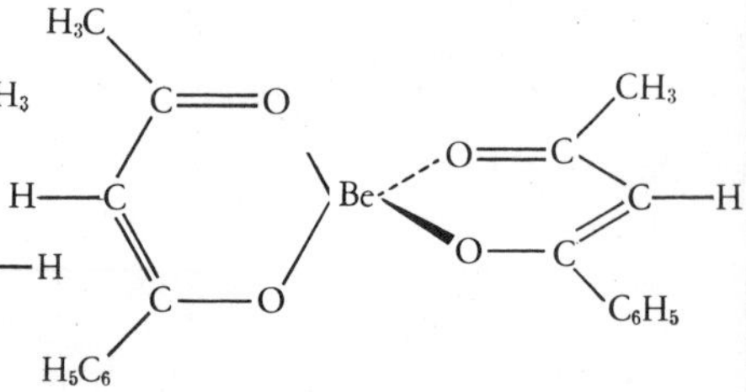

XVII

around the central atom are not required for optical activity; the only requirement is that the molecule and its mirror image be different.

Square planar complexes are very seldom optically active. In most cases (for example, complexes of the type [MABCD]) the plane of the molecule is a plane of symmetry.

Unlike four-coordinated systems, six-coordinated complexes afford many examples of optical isomerism. These are very common among compounds or ions of the type $[M(AA)_3]$. For example, the optical isomers of trioxalatochromate(III) are XVIII and XIX.

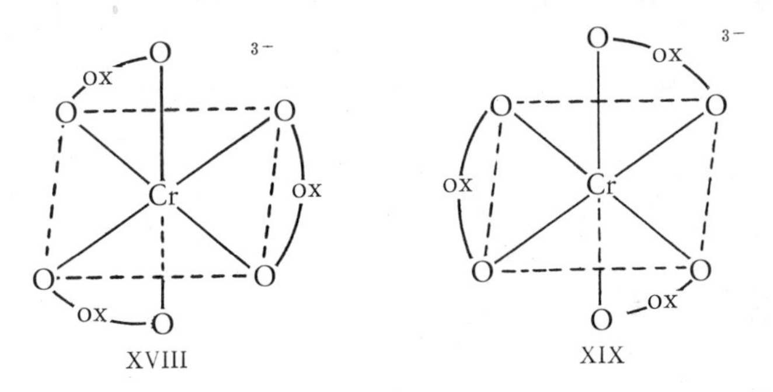

XVIII XIX

Bidentate ligands generally contain carbon, but at least three optically active, purely inorganic complexes are known. One of them was prepared by Werner in order to show that the optical activity in these systems was not due to the presence of carbon. He was able to demonstrate this by using the bridged complex XX, in which the dihydroxo complex XXI is a bidentate ligand. The fact that com-

H
O
Co Co$(NH_3)_4$
O
H
$6+$, 3

XX

H
O
Co$(NH_3)_4$
O
H
$+$

XXI

plexes of the type $[M(AA)_3]$ can be resolved into optical isomers is good evidence that these complexes have the octahedral configuration. Neither a trigonal prism or a planar structure would give rise to optical activity (Table 1-5).

Another very common type of optically active complex has the general formula $[M(AA)_2X_2]$. In this system it is important to note that the trans isomer has a plane of symmetry and cannot be optically active. Therefore, the cis structure for such a complex is conclusively demonstrated if the complex is shown to be optically active. This technique for proof of structure has often been used; the identity of the cis and trans isomers of the new complex dichlorobis(ethylenediamine)rhodium(III), XXII, XXIII, and XXIV,

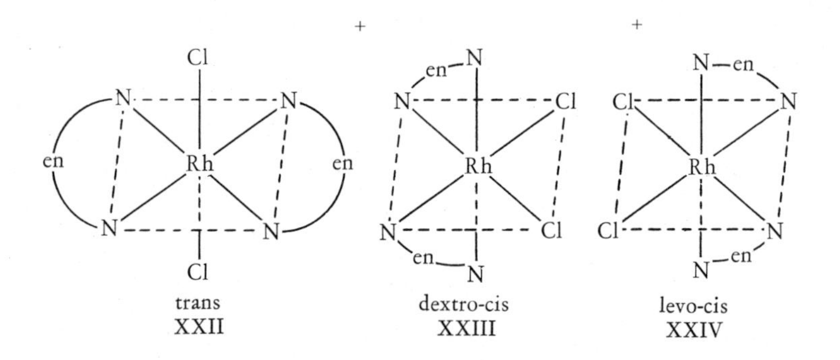

trans XXII

dextro-cis XXIII

levo-cis XXIV

was determined by this technique. One of the isomers of $[Co(en)(NH_3)_2Cl_2]^+$ can exist in nonidentical mirror image forms. These were obtained, XXV and XXVI, and used as proof of the cis-cis structure of the complex.

dextro XXV

levo XXVI

Many examples of this type are known for platinum(IV) complexes.

Multidentate ligands can also give rise to optical isomerism in metal complexes. One of the many such cases is that of *d* and *l* $-[Co(EDTA)]^-$, XXVII and XXVIII.

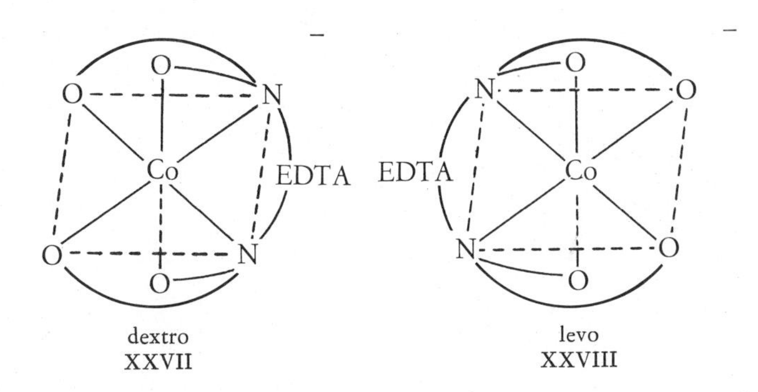

dextro
XXVII

levo
XXVIII

In none of the above examples is the optical activity due to the presence of six different ligands around the central atom. In a complex containing six different ligands, the central atom is asymmetrically coordinated; each of its fifteen geometrical isomers should be resolvable into optical isomers. Thus for one form of $[Pt(py)(NH_3)(NO_2)(Cl)(Br)(I)]$ the optical isomers are XXIX and XXX.

Br
py NO_2
Pt
Cl NH_3
I

Br
O_2N py
Pt
H_3N Cl
I

dextro
XXIX

levo
XXX

However, the resolution of a complex of this type has not yet been achieved.

In conclusion it should be noted that the designation of an optical isomer as either dextro or levo is meaningful only if the wavelength of the light used is known. That an optical isomer may rotate the plane of polarized light to the right (dextro) at one

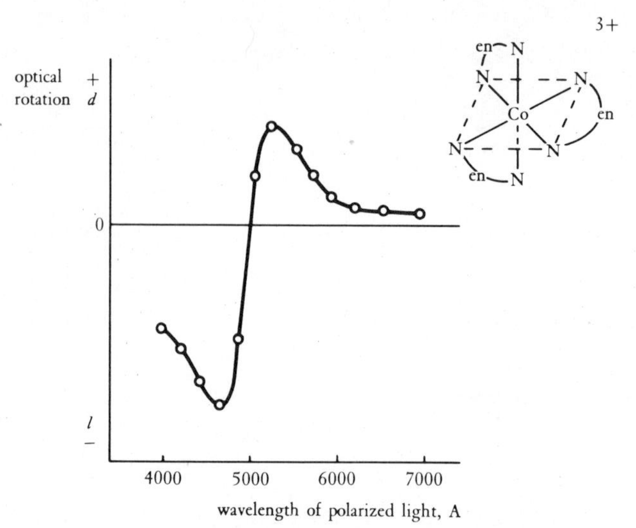

Figure 3-3 The rotatory dispersion curves and the structures for the optical isomers of $[Co(en)_3]^{3+}$.

wavelength but to the left at another is clearly shown in Figure 3-3. The mirror image isomer gives the mirror image curve. Such plots of optical rotation vs. wavelength of light are called *rotatory dispersion curves*. They are more meaningful and useful than just the optical rotation at one wavelength. The absolute configuration of $(+)_{Na}-[Co(en)_3]^{3+}$ was determined by means of X-ray diffraction studies. Using this as a standard it has been possible to assign absolute structures to other complexes by comparison of their rotatory dispersion curves.

3-5 OTHER TYPES OF ISOMERISM

Several types of isomerism other than geometrical and optical are known for coordination compounds. These are often unique to this class of compound. Specific examples are given to represent each type. In general, the nature of the isomerism is sufficiently obvious from the examples that no lengthy discussion is required.

Coordination Isomerism

Compounds containing both cationic and anionic complexes are capable of forming isomers of the coordination isomer type whenever two different, random combinations are possible between and including the extremes $[MA_n][M'X_m]$ and $[M'A_m][MX_n]$. Some examples are

$[Co(NH_3)_6][Cr(C_2O_4)_3]$ and $[Cr(NH_3)_6][Co(C_2O_4)_3]$
$[Rh(en)_3][IrCl_6]$, $[Rh(en)_2Cl_2][Ir(en)Cl_4]$, and $[Ir(en)_3][RhCl_6]$
$[Pt(II)(NH_3)_4][Pt(IV)Cl_6]$ and $[Pt(IV)(NH_3)_4Cl_2][Pt(II)Cl_4]$
$[Cr(NH_3)_6][Cr(NCS)_6]$ and $[Cr(NH_3)_4(NCS)_2][Cr(NH_3)_2(NCS)_4]$

A special type of coordination isomerism is one that involves the different placement of ligands in a bridged complex. This is sometimes called *coordination position isomerism*, and a specific example is provided by the isomers

```
              H
              O
             / \
[(NH3)4Co         Co(NH3)2Cl2]SO4
             \ /
              O
              H
```

and

```
               H
               O
              / \
[Cl(NH3)3Co         Co(NH3)3Cl]SO4
              \ /
               O
               H
```

Ionization Isomerism

This name is used to describe isomers that yield different ions in solution. A classical example is the purple $[Co(NH_3)_5Br]SO_4$ and red $[Co(NH_3)_5SO_4]Br$, which give sulfate and bromide ions, respectively, in solution. Two sets of the many isomers of this type are

$[Co(en)_2(NCS)_2]Cl$ and $[Co(en)_2(NCS)Cl]NCS$
$[Pt(NH_3)_3Br]NO_2$ and $[Pt(NH_3)_3NO_2]Br$

Very similar to these are the isomers resulting from replacement of a coordinated group by water of hydration. This type of isomerism is sometimes called *hydration isomerism.* The best known example is the trio of compounds $[Cr(H_2O)_6]Cl_3$, $[Cr(H_2O)_5Cl]Cl_2 \cdot H_2O$, and $[Cr(H_2O)_4Cl_2]Cl \cdot 2H_2O$, which contain six, five, and four coordinated water molecules, respectively. These isomers differ markedly in physical and chemical properties. Other isomers of the same type are

$[Co(en)_2(H_2O)Cl]Cl_2$ and $[Co(en)_2Cl_2]Cl \cdot H_2O$
$[Cr(py)_2(H_2O)_2Cl_2]Cl$ and $[Cr(py)_2(H_2O)Cl_3] \cdot H_2O$

Linkage Isomerism

Isomerism of the linkage type may result whenever a monodentate ligand has two different atoms available for coordination. The linkage between the metal and ligand in one isomer is through one ligand atom, and that of its isomer is through another. It has been known for many years that nitrite ion in cobalt(III) complexes can be attached either through the nitrogen, $Co—NO_2$ (nitro), or the oxygen, Co—ONO (nitrito). The nitrito complexes of cobalt(III) are unstable and rearrange to form the more stable nitro-isomers. Recent studies have shown that similar linkage isomers can be obtained in complexes of Rh(III), Ir(III), and Pt(IV). Some examples of this type of isomerism are

$[(NH_3)_5Co—NO_2]Cl_2$ and $[(NH_3)_5Co—ONO]Cl_2$

$[(NH_3)_2(py)_2Co(—NO_2)_2]NO_3$ and $[(NH_3)_2(py)_2Co(—ONO)_2]NO_3$

$[(NH_3)_5Ir—NO_2]Cl_2$ and $[(NH_3)_5Ir—ONO]Cl_2$

All ligands other than NO_2^- are written to the left of the metal in order to emphasize how the nitrite ion is bound to the metal.

Many other ligands are potentially capable of forming linkage isomers. Theoretically, all that is required is that two different atoms of the ligand contain an unshared electron pair. Thus the thiocyanate ion, $:N:::C:\ddot{\underset{\cdot\cdot}{S}}:^-$, can attach itself to the metal either through the nitrogen, M—NCS, or the sulfur, M—SCN. Each type of attachment does occur, but generally to give only one form or the other in any particular system. Usually the first-row transition elements are attached through nitrogen, whereas the second- and third-row transition elements (in particular the platinum metals) are attached through sulfur. Very recently it was possible to prepare the following linkage isomers of this type:

$[\{(C_6H_5)_3P\}_2Pd(\text{—}SCN)_2]$ and $[\{(C_6H_5)_3P\}_2Pd(\text{—}NCS)_2]$
$[(OC)_5Mn\text{—}SCN]$ and $[(OC)_5Mn\text{—}NCS]$

Other ligands that should be capable of forming linkage isomers are

$:C{\equiv}O:$ $\quad$ $:C{\equiv}N:^-$ $\quad$ $[\ddot{O}\ \ddot{O}\ S\ \ddot{S}\ \ddot{O}]^{=}$ $\quad$ $H_2\ddot{N}\ C{=}\ddot{O}\ H_2\ddot{N}$

$H_2\ddot{N}\ C{=}\ddot{S}\ H_2\ddot{N}$ $\quad$ $H_3C\ \ddot{O}:\ \dot{S}\ H_3C$

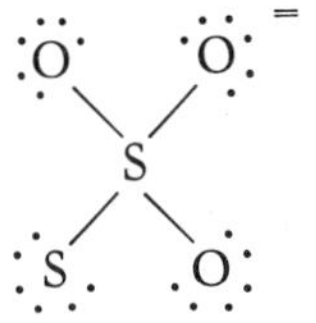

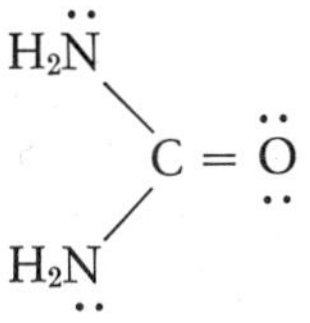

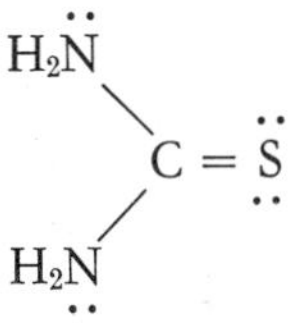

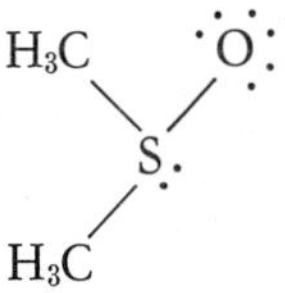

In metal carbonyls as in cyanide complexes the bonding is always metal-carbon.

PROBLEMS

1. Predict the geometry of the following ions:

$[Co(CN)_6]^{3-}$ (diamagnetic); $[NiF_6]^{4-}$ (two unpaired electrons); $[CrF_6]^{4-}$ (four unpaired electrons); $[AuCl_4]^-$ (diamagnetic); $[FeCl_4]^-$ (five unpaired electrons); $[NiF_6]^{2-}$ (diamagnetic).

2. Draw all of the possible isomers of each of the following:

$[Co(NH_3)_4Cl_2]^+$ $\quad$ $[Be(gly)_2]$
$[Rh(en)_2Br_2]^+$ $\quad$ $[Pt(en)Br_2Cl_2]$
$[Ir(C_2O_4)_2Cl_2]^{3-}$ $\quad$ $[Co_2(NH_3)_6(OH)_2Cl_2]^{2+}$

$[Cr(gly)_3]$ $[Pt_2\{P(C_2H_5)_3\}_2Cl_4]$
$[Pt(gly)_2]$ $[Cr(EDTA)]^-$

3. (*a*) A complex $[M(AB)_2]$ is known to be optically active. What does this indicate about the structure of the complex? (*b*) A complex of the type $[M(AA)_2X_2]$ is known to be optically active. What does this indicate about the structure of the complex?

REFERENCES

See also the references given at the end of Chapter I.

A. F. Wells, *Structural Inorganic Chemistry*, 3d. ed., Oxford, Fair Lawn, N.J., 1962.

J. C. Bailar, Jr., "The numbers and structures of isomers of hexacovalent complexes," *J. Chem. Educ.*, **34,** 334 (1957).

IV

Preparations and Reactions of Coordination Compounds

A most important part of chemistry has always been the preparation of compounds. Certainly research in the chemical industry is largely oriented toward the synthesis of new and useful materials. The chemist is very much interested in preparing new compounds, because that is an excellent way of increasing our knowledge of chemistry. Chapter I relates how the synthesis of the first coordination compounds led to the development of concepts and theories that are of considerable value today. The recent preparation of XeF_4 is another example of a synthesis that has led to a tremendous amount of research effort in both synthetic and theoretical chemistry.

In this chapter it is convenient to divide metal coordination compounds into two groups: (1) *Werner complexes* and (2) *metal carbonyls* and *organometallic compounds*. The classification places all complexes that do not contain a metal-carbon bond and all cyanide complexes in group 1. These Werner complexes are the type frequently encountered in the scheme of qualitative analysis for metal ions. Group 2 includes compounds that contain at least one metal-carbon bond. Unlike compounds in group 1, which normally have saltlike properties, the members of group 2 are usually covalent molecular materials. Thus they are generally soluble in nonpolar solvents and have relatively low melting and boiling points. Included in this class are the metal carbonyls and other metal-carbon bonded systems, the *organometallic compounds*, for example, $Hg(C_2H_5)_2$, $K[Pt(C_2H_4)Cl_3]$, and $Fe(C_5H_5)_2$.

Several different but related experimental methods can be used to prepare metal complexes. Some of them are described below; specific examples are given for each of them. The method of choice depends upon the system in question, and not all methods are necessarily applicable to the synthesis of a particular compound. Finding a reaction that produces the desired compound in good yield is only the beginning. The next step is to find a suitable way to isolate the product from its reaction mixture. For the compounds of group 1 this is generally some means of crystallization. Several techniques are available, but among the more commonly used are the following:

1. Evaporate the solvent and cool the more concentrated reaction mixture in an ice-salt bath. Adding a seed crystal of the desired compound and scratching the inside of the beaker below the liquid surface often help to induce crystallization.

2. Slowly add a solvent that is miscible with the solvent of the reaction mixture but that does not dissolve the desired compound. The techniques of cooling, seeding, and scratching may be necessary to cause the product to precipitate from the mixed solvent in which it is insoluble.

3. If the desired complex is a cation, it may be isolated by the addition of an appropriate anion with which it forms an insoluble salt. A suitable cation may be added to the reaction mixture to precipitate an anionic complex.

Compounds in group 2 are also sometimes isolated by these same techniques. In addition, they can also be collected and purified by distillation, sublimation, and chromatographic processes.

4–1 SUBSTITUTION REACTIONS IN AQUEOUS SOLUTION

The substitution reaction in aqueous solution is by far the most common method used for the synthesis of metal complexes. The method involves a reaction between a metal salt in water solution and a coordinating agent. For example, the complex $[Cu(NH_3)_4]SO_4$ is readily prepared by the reaction between an aqueous solution of $CuSO_4$ and excess NH_3 (1). That the coordinated water is instantly

$$\underset{\text{blue}}{[Cu(H_2O)_4]^{2+}} + 4NH_3 \rightarrow \underset{\text{dark blue}}{[Cu(NH_3)_4]^{2+}} + 4H_2O \qquad (1)$$

replaced by ammonia at room temperature is indicated by the change in color from light blue to dark blue. The dark-blue salt crystallizes from the reaction mixture upon the addition of ethanol.

Substitution reactions of metal complexes can also be fairly slow, and for such systems more drastic experimental conditions are required. To prepare $K_3[Rh(C_2O_4)_3]$, one must boil a concentrated aqueous solution of $K_3[RhCl_6]$ and $K_2C_2O_4$ for 2 hr and then evaporate until the product crystallizes from the solution (2).

$$\underset{\text{wine red}}{K_3[RhCl_6]} + 3K_2C_2O_4 \xrightarrow[100^\circ]{\substack{H_2O \\ 2\text{ hr}}} \underset{\text{yellow}}{K_3[Rh(C_2O_4)_3]} + 6KCl \qquad (2)$$

It is also possible that more than one type of ligand will be replaced during a reaction. Thus $[Co(en)_3]Cl_3$ may be prepared by reaction (3). This reaction is rather slow at room temperature;

$$\underset{\text{purple}}{[Co(NH_3)_5Cl]Cl_2} + 3en \rightarrow \underset{\text{orange}}{[Co(en)_3]Cl_3} + 5NH_3 \qquad (3)$$

therefore, it is carried out on a steam bath.

The examples given above are for the preparation of complexes containing only the entering ligand. Such complexes are by far the easiest to prepare, because an excess of coordinating agent may be used to force the equilibrium toward the completely substituted complex. Theoretically, it should be possible to obtain the intermediate mixed complexes, because substitution reactions are known to proceed in a stepwise fashion (Section 5–1). In practice, however, it is often very difficult to isolate the desired mixed complex directly from the reaction mixture. Some successful syntheses of mixed compounds have been achieved by limiting the concentration of a potential ligand. The complex $[Ni(phen)_2(H_2O)_2]Br_2$ can be isolated from a reaction mixture containing two equivalents of phen to one of $NiBr_2$. Similarly, the compound diammineethylenediamineplatinum(II) chloride can be prepared by reactions (4) and (5). Reaction

$$\underset{\text{red}}{K_2[PtCl_4]} + en \rightarrow \underset{\text{yellow}}{[Pt(en)Cl_2]} + 2KCl \tag{4}$$

$$\underset{\text{yellow}}{[Pt(en)Cl_2]} + 2NH_3 \rightarrow \underset{\text{colorless}}{[Pt(en)(NH_3)_2]Cl_2} \tag{5}$$

(4) is successful primarily because the dichloro product is nonionic and separates from the aqueous reaction mixture as it is formed

4–2 SUBSTITUTION REACTIONS IN NONAQUEOUS SOLVENTS

Reactions in solvents other than water had not been used extensively for the preparation of metal complexes until rather recently. Two of the chief reasons why it is sometimes necessary to use solvents other than water are that either (1) the metal ion has a large affinity for water or (2) the ligand is insoluble in water. A few common ions that have a large affinity for water, and thus form strong metal-oxygen bonds, are Al(III), Fe(III), and Cr(III). The addition of basic ligands to aqueous solutions of these metal ions generally results in the formation of a gelatinous hydroxide precipitate rather than a complex containing the added ligands. The metal-oxygen bonds remain intact, but oxygen-hydrogen bonds are broken; the hydrated metal ions behave as protonic acids.

The aqueous reaction between a chromium(III) salt and ethylene diamine is shown by Equation (6). If instead an anhydrous

$$\underset{\text{violet}}{[Cr(H_2O)_6]^{3+}} + 3en \xrightarrow{H_2O} \underset{\text{green}}{[Cr(H_2O)_3(OH)_3]} \downarrow + 3enH^+ \tag{6}$$

chromium salt and a nonaqueous solvent are used, the reaction proceeds smoothly to yield the complex $[Cr(en)_3]^{3+}$ (7). Although

$$\underset{\text{purple}}{CrCl_3} + 3en \xrightarrow{\text{ether}} \underset{\text{yellow}}{[Cr(en)_3]Cl_3} \tag{7}$$

many amminechromium(III) complexes are known, almost never are any of them prepared directly by reaction in water solution. One solvent that has been used rather extensively of late is dimethylformamide (DMF), $(CH_3)_2NCHO$. By using this solvent it was possible to prepare *cis*-$[Cr(en)_2Cl_2]Cl$ in good yield by direct reaction (8).

$$\underset{\text{blue-gray}}{[Cr(DMF)_3Cl_3]} + 2en \xrightarrow{\text{DMF}} \underset{\text{violet}}{cis\text{-}[Cr(en)_2Cl_2]Cl} \tag{8}$$

In some cases a nonaqueous solvent is required because the ligand is not water soluble. Often it suffices to dissolve the ligand in a water-miscible solvent and then to add this solution to a concentrated water solution of the metal ion. Metal complexes of bipy and phen are generally prepared in this way. Thus the addition of an alcoholic solution of bipy to an aqueous solution of $FeCl_2$ readily yields the complex $[Fe(bipy)_3]Cl_2$ (9).

$$\underset{\text{colorless}}{[Fe(H_2O)_6]^{2+}} + 3bipy \xrightarrow{H_2O—C_2H_5OH} \underset{\text{intense red}}{[Fe(bipy)_3]^{2+}} + 6H_2O \tag{9}$$

4-3 SUBSTITUTION REACTIONS IN THE ABSENCE OF SOLVENT

The direct reaction between an anhydrous salt and a liquid ligand can be used to prepare metal complexes. In many cases the liquid ligand present in very large excess also serves as a solvent for the reaction mixture. A method applicable to the synthesis of metal ammines involves the addition of a metal salt to liquid ammonia followed by evaporation to dryness. Evaporation occurs readily at room temperature, because ammonia boils at $-33°C$. The dry residue obtained is essentially the pure metal ammine. For example, $[Ni(NH_3)_6]Cl_2$ can be prepared in this way by reaction (10).

$$\underset{\text{yellow}}{NiCl_2} + 6NH_3(\text{liquid}) \rightarrow \underset{\text{violet}}{[Ni(NH_3)_6]Cl_2} \tag{10}$$

Often, this is not the method of choice, because aqueous ammonia is more convenient to use and generally gives the same result. However, in some cases, as in the preparation of $[Cr(NH_3)_6]Cl_3$, it is necessary to use liquid ammonia to avoid the formation of $Cr(OH)_3$ [see (6)].

One method used for the preparation of $[Pt(en)_2]Cl_2$ or $[Pt(en)_3]Cl_4$ is the direct reaction between ethylenediamine and $PtCl_2$ or $PtCl_4$, respectively. The technique is to add slowly the solid platinum salts to the liquid ethylenediamine. This addition is accompanied by a vigorous evolution of heat, which is to be expected whenever a strong acid is added to a strong base. Recall (Section 2–1) that in terms of the Lewis definition of acids and bases, the formation of coordination compounds involves an acid-base reaction. In this particular case the platinum ions are the acids and ethylenediamine is the base. Recently, numerous metal dimethylsulfoxide complexes have been prepared and characterized. One method used to prepare some of these was a direct reaction in the absence of any added solvent (11).

$$\underset{\text{pink}}{Co(ClO_4)_2} + 6(CH_3)_2SO \rightarrow \underset{\text{pink}}{[Co\{(CH_3)_2SO\}_6](ClO_4)_2} \qquad (11)$$

4–4 THERMAL DISSOCIATION OF SOLID COMPLEXES

Thermal dissociation amounts to a substitution reaction in the solid state. At some elevated temperature, volatile coordinated ligands are lost and their place in the coordination sphere is taken by the anions of the complex. A familiar example, but one that is perhaps seldom considered in this way, is the loss of water by $CuSO_4{\cdot}5H_2O$ when it is heated. The blue hydrate yields the almost white anhydrous sulfate by reaction (12). The replacement of water

$$\underset{\text{blue}}{[Cu(H_2O)_4]SO_4{\cdot}H_2O} \xrightarrow{\Delta} \underset{\text{colorless}}{[CuSO_4]} + 5H_2O \uparrow \qquad (12)$$

ligands by sulfate ions is responsible for the change in color. The hydrated copper(II) ion absorbs light near the infrared end of the visible spectrum; this is responsible for its blue color. Since the crystal field splitting due to sulfate ion is less than that of water, copper(II) ion in a sulfate ion environment absorbs light of a longer wavelength. This places the anhydrous copper sulfate absorption in the infrared and hence the compound has no absorption in the visible spectrum and is colorless. The situation described on page 20 for invisible ink is another illustration of a solid state reaction.

At elevated temperatures coordinated water can often be liberated from aquoamminemetal complexes. It is sometimes convenient to use this method to prepare halogenoamminemetal compounds (13).

$$\underset{\text{colorless}}{[Rh(NH_3)_5H_2O]I_3} \xrightarrow{100^\circ} \underset{\text{yellow}}{[Rh(NH_3)_5I]I_2} + H_2O \uparrow \qquad (13)$$

Just as water can be expelled from solid aquo complexes, ammonia and amines can sometimes be liberated from metal ammines. This procedure is used to prepare acidoamminemetal[1] complexes. This is a general method for the synthesis of compounds of the type *trans*-$[PtA_2X_2]$. The reaction yields the trans isomer, as described in Section 4–8. The most common example of this is the preparation of *trans*-$[Pt(NH_3)_2Cl_2]$ by the thermal evolution of ammonia (14).

$$\underset{\text{white}}{[Pt(NH_3)_4]Cl_2} \xrightarrow{250^\circ} \underset{\text{yellow}}{\textit{trans}\text{-}[Pt(NH_3)_2Cl_2]} + 2NH_3 \uparrow \qquad (14)$$

The corresponding reaction for the analogous pyridine system takes place at approximately a hundred degrees lower temperature. The best method for the synthesis of *trans*-$[Cr(en)_2(NCS)_2]NCS$ is the liberation of ethylenediamine from solid $[Cr(en)_3](NCS)_3$ (15). This reaction gives better results if the starting material contains a small

$$\underset{\text{yellow}}{[Cr(en)_3](NCS)_3} \xrightarrow[NH_4SCN]{130^\circ} \underset{\text{orange}}{\textit{trans}\text{-}[Cr(en)_2(NCS)_2]NCS} + en \uparrow \qquad (15)$$

---

[1] Acido is used as a general term referring to anionic ligands. Likewise, the use of ammine with reference to a general class of compounds is not specific for ammonia but includes other amines as well. Thus $[Co(NH_3)_4Br_2]^+$, $[Cr(en)(C_2O_4)_2]^-$, and $[Pt(py)_2(NO_2)_2]$ are all examples of acidoammine complexes.

amount of ammonium thiocyanate. Such thermal reactions do not necessarily lead to the formation of a trans isomer. Thus if $[Cr(en)_3]Cl_3$ is heated at 210°C, the product obtained is *cis*-$[Cr(en)_2Cl_2]Cl$. It is not yet understood why certain salts in the chromium system give a preferred geometric isomer upon thermal dissociation.

4–5 OXIDATION–REDUCTION REACTIONS

The preparation of many metal complexes often involves an accompanying oxidation-reduction reaction. For the hundreds of cobalt(III) complexes that have been prepared, the starting material was almost always some cobalt(II) salt. This is because the usual oxidation state of cobalt in its simple salts is 2. The oxidation state of 3 becomes the stable form only when cobalt is coordinated to certain types of ligands (Section 5–2). Furthermore, it is convenient to start with salts of cobalt(II) because Co(II) complexes undergo substitution reactions very rapidly, whereas reactions of Co(III) complexes are very slow (Section 6–4). The preparation of Co(III) complexes therefore proceeds by a fast reaction between cobalt(II) and the ligand to form a cobalt(II) complex which is then oxidized to the corresponding cobalt(III) complex. For example, reaction (16) is presumed to involve first the formation of $[Co(NH_3)_6]^{2+}$

$$\underset{\text{pink}}{4[Co(H_2O)_6]Cl_2} + 4NH_4Cl + 20NH_3 + O_2 \rightarrow \underset{\text{orange}}{4[Co(NH_3)_6]Cl_3} + 26H_2O \quad (16)$$

(17) followed by its oxidation (18).

$$\underset{\text{pink}}{[Co(H_2O)_6]Cl_2} + 6NH_3 \rightarrow \underset{\text{rose}}{[Co(NH_3)_6]Cl_2} + 6H_2O \quad (17)$$

$$\underset{\text{rose}}{4[Co(NH_3)_6]Cl_2} + 4NH_4Cl + O_2 \rightarrow \underset{\text{orange}}{4[Co(NH_3)_6]Cl_3} + 4NH_3 + 2H_2O \quad (18)$$

Although air oxidation is commonly used in the synthesis of cobalt(III) complexes, other oxidizing agents can be employed.

Many oxidizing agents are able to oxidize Co(II) to Co(III) in the presence of suitable ligands; only a few are convenient to use. Oxidizing agents such as potassium permanganate and potassium dichromate introduce to the reaction mixture ions that are not easily separated from the desired product. Oxidizing agents such as oxygen and hydrogen peroxide do not introduce foreign metal ions to the reaction mixture. Another type of suitable oxidizing agent is one whose reduction product is insoluble and can be removed by filtration. This is true of PbO_2, which is reduced to Pb^{2+} and can be removed as insoluble $PbCl_2$. Similarly, SeO_2 yields insoluble Se.

It is of interest to note that the reaction product may sometimes depend upon the particular oxidizing agent employed. For example, $[Co(EDTA)]^-$ is prepared by the oxidization of $[Co(EDTA)]^{2-}$ with $[Fe(CN)_6]^{3-}$. If Br_2 is used as the oxidizing agent, the reaction product is $[Co(EDTA)Br]^{2-}$. This difference results because the first reaction proceeds by the transfer of an electron from the reducing agent to the oxidizing agent (19). The second reaction is believed to in-

$$\underset{\text{pink}}{[Co(EDTA)]^{2-}} + [Fe(CN)_6]^{3-} \rightarrow \underset{\text{violet}}{[Co(EDTA)]^-} + [Fe(CN)_6]^{4-} \qquad (19)$$

volve a direct attack by bromine on cobalt and a bromine atom transfer (20). For a more complete discussion of oxidation-reduction

$$\underset{\text{pink}}{[Co(EDTA)]^{2-}} + Br_2 \rightarrow [(EDTA)Co\text{---}Br_2]^{2-} \rightarrow \underset{\text{rose}}{[(EDTA)Co\text{—}Br]^{2-}} + Br \qquad (20)$$

reactions of metal complexes see Section 6–8.

Less common than preparation of complexes by the oxidation of the central metal ion is preparation by reduction to complexes of the metal ion in a lower oxidation state. One reason that the latter has not been extensively applied is that the resulting compounds are often so sensitive to oxidation that they must be handled in an inert, oxygen- and moisture-free atmosphere. However, with special precautions it is possible to prepare many interesting complexes in which the central metal ion has an unusually low oxidation state. Reductions in liquid ammonia have been useful for this purpose, as is illustrated by reaction (21). The oxidation state of nickel in this compound is zero. The compound is readily oxidized in air, and it

$$\underset{\text{yellow}}{K_2[Ni(CN)_4]} + 2K \xrightarrow[\text{ammonia}]{\text{liquid}} \underset{\text{yellow}}{K_4[Ni(CN)_4]} \qquad (21)$$

reduces water with the liberation of hydrogen. In a few cases it has actually been possible to reduce the central metal ion of a complex to a negative oxidation state. Iron has an oxidation state of 2− in $K_2[Fe(CO)_4]$, which is prepared by reaction (22). The salt is stable

$$\underset{\text{yellow}}{Fe(CO)_5} + 4KOH \rightarrow \underset{\text{colorless}}{K_2[Fe(CO)_4]} + K_2CO_3 + 2H_2O \qquad (22)$$

in aqueous alkaline solution, but it is very sensitive to air oxidation. Still another example of a complex containing a metal ion in a negative oxidation state is $[V(bipy)_3]^-$, which is prepared by the reduction of $[V(bipy)_3]^{3+}$. It is of interest to note that in all of these cases of unusually low oxidation states, the EAN (Section 2–2) of the metal ion is the same as that of the next rare gas.

4–6 CATALYSIS

In systems that react slowly it is often necessary to employ elevated temperatures and long reaction times in order to prepare desired coordination compounds. Alternatively, a catalyst may be used to increase the speed of a reaction. Catalysis has been used successfully in a few instances to prepare metal complexes. Remember that there are two types of catalysis: *heterogeneous catalysis* takes place when the catalyst is in a different phase than that of the reactants; *homogeneous catalysis* occurs when the catalyst and reacting materials are in the same phase. Examples of the use of heterogeneous and homogeneous catalysis in the synthesis of metal complexes are noted below.

The best known example of heterogeneous catalysis in such systems is the preparation of $[Co(NH_3)_6]Cl_3$. It is now recognized that reactions of cobalt(III) complexes are catalyzed by certain solid surfaces such as decolorizing charcoal. For example, an aqueous solution of $[Co(NH_3)_6]Cl_3$ can be boiled for hours without any noticeable change in its yellow-orange color; this indicates no significant amount of reaction. The same treatment with added decolorizing charcoal soon yields a red solution due to the presence of $[Co(NH_3)_5OH_2]^{3+}$. Prolonged heating results in the total destruction of the complex and precipitation of cobalt(II) hydroxide.

The rapid decomposition of $[Co(NH_3)_6]^{3+}$ in water containing decolorizing charcoal suggests that the compound might be formed rapidly in a reaction mixture containing charcoal and excess ammonia. In fact, the air oxidation of a reaction mixture of aqueous cobalt(II) chloride, excess ammonia, and ammonium chloride, followed by acidification with excess hydrochloric acid, yields largely $[Co(NH_3)_5Cl]Cl_2$ (23). Under the same conditions in the presence

$$\underset{\text{pink}}{[Co(H_2O)_6]Cl_2} \xrightarrow[NH_4Cl]{NH_3\text{-}H_2O\text{-}O_2} \xrightarrow{HCl} \underset{\text{purple}}{[Co(NH_3)_5Cl]Cl_2} \qquad (23)$$

of charcoal the product is almost exclusively $[Co(NH_3)_6]Cl_3$ (24).

$$\underset{\text{pink}}{[Co(H_2O)_6]Cl_2} \xrightarrow[NH_4Cl\text{-charcoal}]{NH_3\text{-}H_2O\text{-}O_2} \xrightarrow{HCl} \underset{\text{orange}}{[Co(NH_3)_6]Cl_3} \qquad (24)$$

Cobalt(II) salts in equilibrium with excess ammonia (25) yield primarily the hexaammine cobalt(II) complex.

$$[Co(H_2O)_6]^{2+} \rightleftharpoons [Co(NH_3)(H_2O)_5]^{2+} \rightleftharpoons [Co(NH_3)_2(H_2O)_4]^{2+} \rightleftharpoons [Co(NH_3)_3(H_2O)_3]^{2+} \rightleftharpoons [Co(NH_3)_4(H_2O)_2]^{2+} \rightleftharpoons [Co(NH_3)_5(H_2O)]^{2+} \rightleftharpoons [Co(NH_3)_6]^{2+} \qquad (25)$$

Since the catalyst cannot change the position of the Co(II)-ammonia equilibrium, why is the oxidation product (23) not the hexaammine in the absence of charcoal? An explanation can be offered in terms of our present views on the mechanisms of reaction in these systems. One process for the oxidation of a metal complex seems to require the formation of an activated bridged intermediate and the conduction of electrons through the bridging group (Section 6–8). The air oxidation of $[Co(NH_3)_6]^{2+}$ seems to proceed through such an intermediate. A molecule of oxygen may add to two of the reactive $[Co(NH_3)_6]^{2+}$ cations to form a peroxo-bridged cobalt(III) species (26).

$$\underset{\text{orange}}{2[Co(NH_3)_6]^{2+}} + O_2 \rightarrow \underset{\text{pink}}{[(NH_3)_5Co{-}O{-}O{-}Co(NH_3)_5]^{4+}} + 2NH_3 \quad (26)$$

Bridged complexes of this type are known. It is then necessary that the bridged system react with ammonia to form $[Co(NH_3)_6]^{3+}$; alternatively, the product may result from the reaction of ammonia with $[Co(NH_3)_5OH]^{2+}$, which is generated by cleavage of the O—O bond in the bridged complex (26). In either case the reaction with ammonia must be very slow; however, it occurs smoothly in the presence of decolorizing charcoal (27). In the absence of a catalyst

$$\underset{\text{pink}}{[Co(NH_3)_5OH]^{2+}} \xrightarrow[\text{C, fast}]{NH_3\ \text{very slow}} \underset{\text{orange}}{[Co(NH_3)_6]^{3+}} \quad (27)$$

the reaction is so slow that it effectively does not occur. The observed product in this case, $[Co(NH_3)_5Cl]Cl_2$, is presumably formed from the reaction of HCl with $[(NH_3)_5Co{-}O{-}O{-}Co(NH_3)_5]^{4+}$ (28).

$$\underset{\text{pink}}{[(NH_3)_5Co{-}O{-}O{-}Co(NH_3)_5]^{4+}} \xrightarrow{HCl} \underset{\text{pink}}{[(NH_3)_5Co{-}OH_2]^{3+}} \xrightarrow{HCl} \underset{\text{purple}}{[Co(NH_3)_5Cl]^{2+}} \quad (28)$$

Homogeneous catalysis was recently observed and investigated for the reactions of several platinum(IV) complexes. Such complexes generally react extremely slowly, but in the presence of catalytic amounts of platinum(II), reaction occurs readily without need for drastic experimental conditions. Platinum(II) catalysis has been used successfully to prepare new compounds of platinum(IV) as well as complexes previously obtained by other methods. Complexes of the type *trans*-$[PtA_4X_2]^{2+}$ are generally prepared by the oxidation of $[PtA_4]^{2+}$ with X_2.

Another method now available is the reaction of *trans*-

$[PtA_4Y_2]^{2+}$ with excess X^- in the presence of catalytic amounts of $[PtA_4]^{2+}$. An example is the reaction between *trans*-$[Pt(NH_3)_4Cl_2]^{2+}$ and Br^- in the presence of $[Pt(NH_3)_4]^{2+}$, (29). This method of syn-

$$\underset{\text{yellow}}{trans\text{-}[Pt(NH_3)_4Cl_2]^{2+}} + 2Br^- \xrightarrow{[Pt(NH_3)_4]^{2+}} \underset{\text{orange}}{trans\text{-}[Pt(NH_3)_4Br_2]^{2+}} + 2Cl^- \quad (29)$$

thesis of *trans*-$[Pt(NH_3)_4Br_2]^{2+}$ has no advantage over preparation by the oxidation of $[Pt(NH_3)_4]^{2+}$ with Br_2. However, the analogous thiocyanato complex, *trans*-$[Pt(NH_3)_4(SCN)_2]^{2+}$, had not been prepared until it was made by this method of platinum(II) catalysis (30).

$$\underset{\text{yellow}}{trans\text{-}[Pt(NH_3)_4Cl_2]^{2+}} + 2SCN^- \xrightarrow{[Pt(NH_3)_4]^{2+}} \underset{\text{orange}}{trans\text{-}[Pt(NH_3)_4(SCN)_2]^{2+}} + 2Cl^- \quad (30)$$

Platinum(II) catalysis in these systems is believed to proceed by a mechanism that involves an activated bridged complex and a two-electron redox reaction. Such a process is represented by the reaction scheme (31) to (34). In reaction (31), the catalyst

$$\left[\overset{(NH_3)_4}{Pt}\right]^{2+} + Br^- \rightleftharpoons \left[\overset{(NH_3)_4}{Pt}\text{—}Br\right]^{+} \quad (31)$$

$$\left[Cl\text{—}\overset{(NH_3)_4}{Pt}\text{—}Cl\right]^{2+} + \left[\overset{(NH_3)_4}{Pt}\text{—}Br\right]^{+} \rightleftharpoons \left[Cl\text{—}\overset{(NH_3)_4}{Pt}\text{—}Cl\text{—}\overset{(NH_3)_4}{Pt}\text{—}Br\right]^{3+} \quad (32)$$

$$\left[Cl\text{—}\overset{(NH_3)_4}{Pt}\text{—}Cl\text{—}\overset{(NH_3)_4}{Pt}\text{—}Br\right]^{3+} \rightleftharpoons \left[Cl\text{—}\overset{(NH_3)_4}{Pt}\right]^{+} + \left[Cl\text{—}\overset{(NH_3)_4}{Pt}\text{—}Br\right]^{2+} \quad (33)$$

$$\left[Cl\text{—}\overset{(NH_3)_4}{Pt}\right]^{+} \rightleftharpoons \left[\overset{(NH_3)_4}{Pt}\right]^{2+} + Cl^- \quad (34)$$

$[Pt(NH_3)_4]^{2+}$ is associated weakly with Br^-, which is present in large excess. Recall that evidence was cited in Section 3–1 for the coordination of a fifth and sixth group above and below the square plane of a four-coordinated planar complex.

Equation (32) represents the formation of a bridged complex between platinum(II) and platinum(IV). The transfer of two electrons from platinum(II) to platinum(IV) through the chloride bridging atom results in the original platinum(II) becoming platinum(IV). Since the new Pt(IV) species contains the bromide ion, the reaction gives *trans*-$[Pt(NH_3)_4BrCl]^{2+}$. The reader can repeat this process and see that a similar procedure can lead to the formation of *trans*-$[Pt(NH_3)_4Br_2]^{2+}$. It should also be noted that the catalyst $[Pt(NH_3)_4]^{2+}$ is regenerated in (33) and (34). This mechanism requires that there be platinum exchange between the Pt(II) and Pt(IV) species. Such exchange has in fact been demonstrated by using radioactive platinum as a tracer.

4–7 SUBSTITUTION REACTIONS WITHOUT METAL–LIGAND BOND CLEAVAGE

The formation of some metal complexes has been found to occur without the breakage of a metal-ligand bond. In the preparation of $[Co(NH_3)_5OH_2]^{3+}$ salts from $[Co(NH_3)_5CO_3]^+$, CO_2 is produced by cleavage of a carbon-oxygen bond which leaves the metal-oxygen bond intact (35). This was demonstrated by running the reaction

$$\underset{\text{pink}}{[(NH_3)_5Co{-}O}\,\vdots\,CO_2]^+ + 2H^+ \rightarrow \underset{\text{pink}}{[(NH_3)_5Co{-}OH_2]^{3+}} + CO_2 \quad (35)$$

in ^{18}O-labeled water. Both products contained oxygen with normal isotopic distribution. This clearly indicates that solvent water does not contribute oxygen to the products, and hence the oxygen must come from the reactants. A simple but not conclusive piece of evidence for the preservation of the Co—O bond is the fact that the reaction is complete soon after the acidification of the carbonato compound. Since reactions that involve the breaking of Co—O bonds in a variety of compounds are known to be slow, the fast reaction suggests that another mechanism is involved in this case.

Reactions similar to (35) are believed to be fairly general, and they have been used to prepare aquo complexes from the corresponding carbonato complexes. Other systems such as

$[(NH_3)_5Co—OSO_2]^+$ and $[(NH_3)_5Co—ONO]^{2+}$ may react in a similar fashion to produce $[(NH_3)_5Co—OH_2]^{3+}$ with the liberation of SO_2 and NO, respectively.

The reverse of these processes, reaction (36), may also be expected to occur. Reaction (36) has been studied in detail, and it is

$$\underset{\text{pink}}{[(NH_3)_5Co—{}^{18}OH]^{2+}} + N_2O_3 \rightarrow \underset{\text{pink}}{[(NH_3)_5Co—{}^{18}ONO]^{2+}} + HNO_2 \quad (36)$$

known to occur without Co—O bond cleavage. The best evidence for retention of the Co—O bond is the observation that when the $[Co(NH_3)_5OH]^{2+}$ was labeled with ^{18}O, the product $[Co(NH_3)_5{}^{18}ONO]^{2+}$ contained 99.4 per cent of the ^{18}O originally present in the starting material. It is reasonable to expect a similar behavior for reactions of hydroxo complexes with other acid anhydrides, for example, CO_2 and SO_2. Indeed, $[Co(NH_3)_5CO_3]^+$ can be prepared by the reaction of $[Co(NH_3)_5OH]^{2+}$ with CO_2.

A variety of other reactions occur without metal-ligand bond cleavage. Reactions of this type have been used to convert nitrogen-containing ligands to ammonia. Examples include the oxidation of N-bonded thiocyanate (37) and the reduction of N-bonded nitrite (38). Coordinated ligands often exhibit their own characteristic

$$\underset{\text{orange}}{[(NH_3)_5Co—NCS]^{2+}} \xrightarrow[H_2O]{H_2O_2} \underset{\text{orange}}{[(NH_3)_5Co—NH_3]^{3+}} \quad (37)$$

$$\underset{\text{white}}{[(NH_3)_3Pt—NO_2]^+} \xrightarrow[HCl\text{-}H_2O]{Zn} \underset{\text{white}}{[(NH_3)_3Pt—NH_3]^{2+}} \quad (38)$$

reactions. For example, the hydrolysis (39) and fluorination (40) of PCl_3 can be accomplished while PCl_3 itself functions as a ligand.

$$\underset{\text{yellow}}{[Cl_2Pt(PCl_3)_2]} + 6H_2O \rightarrow \underset{\text{yellow}}{[Cl_2Pt(P(OH)_3)_2]} + 6HCl \quad (39)$$

$$\underset{\text{yellow}}{[Cl_2Pt(PCl_3)_2]} + 2SbF_3 \rightarrow \underset{\text{colorless}}{[Cl_2Pt(PF_3)_2]} + 2SbCl_3 \quad (40)$$

A type of reaction of considerable current interest involves the addition to or substitution on organic molecules coordinated to metal ions. Acetylacetone (acac) (41) forms very stable chelate compounds with a number of metal ions. Both complexes $[Cr(acac)_3]$

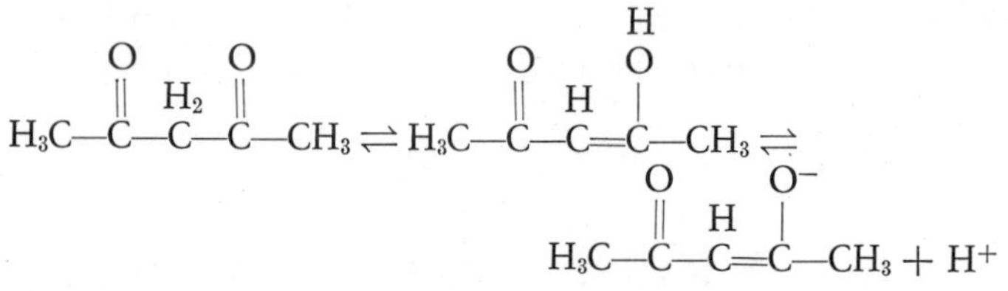

$$H_3C-\overset{\overset{O}{\|}}{C}-\overset{H_2}{C}-\overset{\overset{O}{\|}}{C}-CH_3 \rightleftharpoons H_3C-\overset{\overset{O}{\|}}{C}-\overset{H}{C}=\overset{\overset{\overset{H}{O}}{|}}{C}-CH_3 \rightleftharpoons$$

$$H_3C-\overset{\overset{O}{\|}}{C}-\overset{H}{C}=\overset{\overset{O^-}{|}}{C}-CH_3 + H^+ \quad (41)$$

and $[Co(acac)_3]$ are stable (the equilibrium $[M(H_2O)_6]^{3+} + 3acac \rightleftharpoons [M(acac)_3]$ lies far to the right) and kinetically inert (acac exchange $[M(acac)_3] + acac^* \rightleftharpoons [M(acac)_2(acac^*)]$ is slow). The complex $[Cr(acac)_3]$ reacts rapidly with bromine in glacial acetic acid to yield a chromium chelate in which a hydrogen in each acetylacetone ring has been replaced by a bromine atom (42).

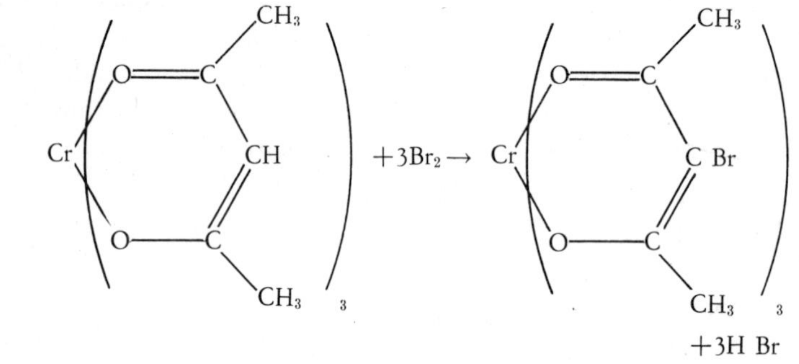

$$\underset{\text{violet}}{\left(Cr\begin{matrix} O=C(CH_3) \\ \quad CH \\ O-C(CH_3) \end{matrix}\right)_3} + 3Br_2 \rightarrow \underset{\text{brown}}{\left(Cr\begin{matrix} O=C(CH_3) \\ \quad C\,Br \\ O-C(CH_3) \end{matrix}\right)_3} + 3H\,Br \quad (42)$$

Similar iodo and nitro compounds, as well as analogous derivatives of other metals, have been prepared.

4-8 TRANS EFFECT

The preparation of square planar compounds of platinum(II) has been put on a somewhat systematic basis by Russian workers, who observed that certain ligands cause the groups across from them in the square plane (trans position) to be replaced easily. Ligands that labilize the group trans to them are said to have a strong trans-directing influence (*trans* effect). The classic example of this is the preparation of the isomeric dichlorodiammineplatinum(II) complexes (43), (44). In sequences (43) and (44), whenever a choice is

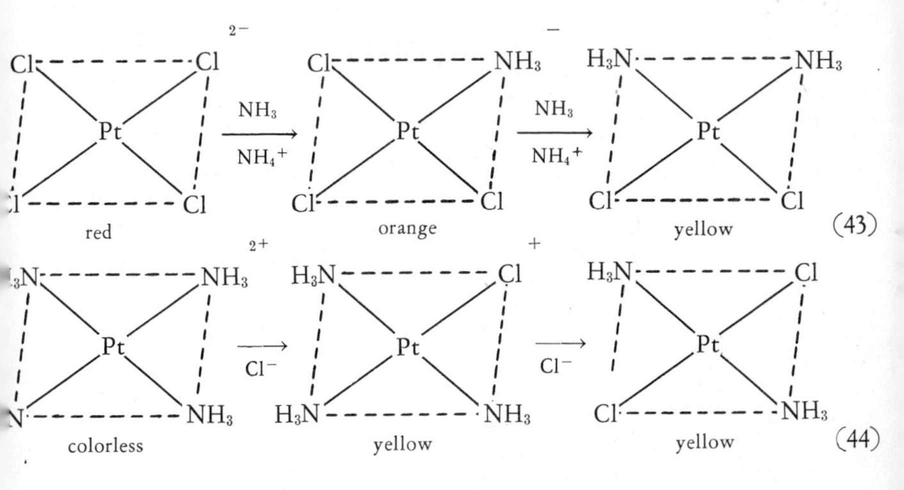

possible, the group opposite to a Cl^- is replaced in preference to one trans to NH_3. On this basis Cl^- is said to have a greater trans effect than NH_3.

After ligands are classified with respect to their trans-directing ability, it is possible to use the information to synthesize specific desired compounds. The three isomers of $[Pt(CH_3NH_2)NH_3(NO_2)Cl]$ can be prepared by the reactions outlined in Figure 4–1. The success of the procedure depends on the existence of this trans effect order: $NO_2^- > Cl^- > NH_3 \sim CH_3NH_2$. However, this information on the trans effect will not of itself permit one to propose such a reaction sequence. Another consideration is the stability of the various platinum-ligand bonds. The trans effect will explain the synthetic results in steps (*a*), (*c*), and (*f*); but the relative ease of replacement of chloride coordinated to platinum(II) accounts for steps (*b*), (*d*), and (*e*).

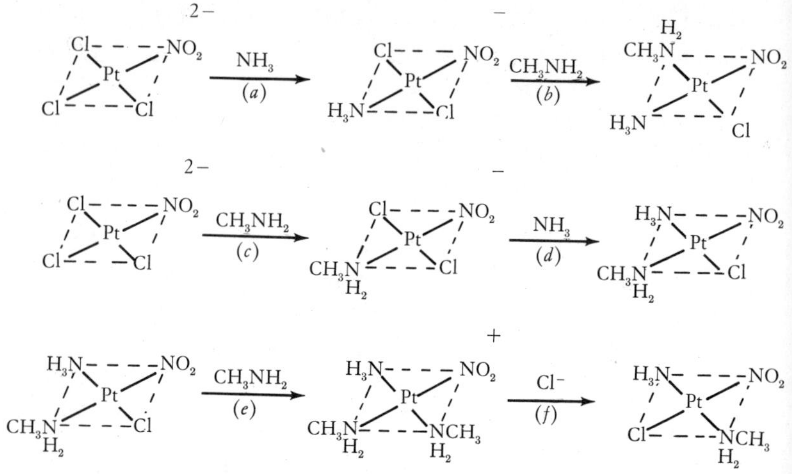

Figure 4–1 The preparation of the three isomers of $[Pt(CH_3NH_2)(NH_3)(NO_2)Cl]$.

Rather extensive studies show that the trans effects of a variety of ligands decrease in the order:

$$CN^- \sim CO \sim C_2H_4 > PH_3 \sim SH_2 > NO_2^- > I^- > Br^- > Cl^- > NH_3 \sim py > OH^- > H_2O$$

This information is now used to design preparations of Pt(II) compounds or to predict their kinetic behavior. Investigations have demonstrated that the trans effect phenomenon is not nearly as important in complexes of other metals.

4–9 SYNTHESIS OF CIS-TRANS ISOMERS

There are two approaches to the preparation of cis-trans isomers: preparation of a mixture of isomers (necessitating subsequent separation) and stereospecific synthesis that yields a single product. The second approach has been most successful for the synthesis of isomers of platinum(II) complexes by taking advantage of the trans effect (Section 4–8). Reactions of cobalt(III) complexes often yield a mixture of cis-trans isomers that then must be separated.

Complexes of platinum(IV) can readily be prepared as the trans isomer, whereas the cis form is much more difficult to obtain. The oxidation of square planar platinum(II) complexes generally yields the corresponding octahedral platinum(IV) complex. The preparation of *trans*-$[Pt(NH_3)_4Cl_2]^{2+}$ is achieved in this manner (45). Oxi-

H_3N, NH_3, Pt^{II}, H_3N, NH_3 $^{2+}$ $+ Cl_2 \xrightarrow[HCl]{H_2O}$ Cl, H_3N, NH_3, Pt^{IV}, H_3N, NH_3, Cl $^{2+}$

colorless yellow (45)

dizing agents such as Br_2 and H_2O_2 yield the trans dibromo and dihydroxo complexes, respectively. Synthesis of *cis*-$[Pt(NH_3)_4Cl_2]Cl_2$ requires the more tedious approach given in (46). The replacement

N, Cl, Pt^{II}, N, NH_3 $^{+}$ $\xrightarrow[HCl]{Cl_2}$ Cl, H_3N, Cl, Pt^{IV}, H_3N, NH_3, Cl $^{+}$ $\xrightarrow{NH_3}$ Cl, H_3N, Cl, Pt^{IV}, H_3N, NH_3, NH_3 $^{2+}$

yellow yellow yellow (46)

of one or more ligands in a particular isomer by other groups is sometimes used to prepare new compounds having a desired stereochemistry. This is not a reliable method, however, because it is known that in many cases the stereochemistry of the starting material is not preserved during reaction.

Preparations of certain cis or trans isomers have been previously discussed: *cis*-$[Cr(en)_2Cl_2]Cl$, page 88; *cis*- and *trans*-$[Pt(NH_3)_2Cl_2]$, page 97; and *trans*-$[Cr(en)_2(NCS)_2]NCS$, page 87. In all of these preparations the product is largely the indicated isomer, but reactions frequently produce mixtures of isomers. These can then be separated by means of fractional crystallization, ion exchange chromatography, or other physical techniques. When KOH is added slowly to a refluxing wine-red aqueous solution of $RhCl_3 \cdot 3H_2O$ and $H_2NCH_2CH_2NH_2 \cdot 2HCl(en \cdot 2HCl)$, one obtains a clear yellow solution. Addition of HNO_3 to the cold reaction mixture yields a golden-yellow crystalline product, which has been shown to be *trans*-$[Rh(en)_2Cl_2]NO_3$. Evaporation of the resulting solution causes the precipitation of the more soluble, bright-yellow *cis*-$[Rh(en)_2Cl_2]NO_3$, Equation (47). This illustrates a reaction that

$$\underset{\text{red}}{RhCl_3 \cdot 3H_2O} + 2en \cdot 2HCl \xrightarrow[\substack{H_2O \\ 100^\circ}]{KOH} \text{yellow solution} \xrightarrow{\text{cool}\ HNO_3}$$

$$\underset{\text{yellow}}{cis\text{-}[Rh(en)_2Cl_2]^+} \xleftarrow[25^\circ]{\text{evaporate}} \text{yellow solution} + \underset{\text{yellow}}{trans\text{-}[Rh(en)_2Cl_2]^+} \quad (47)$$

produces a mixture of isomers that can be readily separated by taking advantage of difference in solubilities.

Since in many, if not most, attempts to produce isomeric materials one obtains a mixture of isomers or a single isomer of unknown stereochemistry, methods for determining structure are required. Several chemical tests for geometric structure are available. An elegant and absolute chemical test is the resolution of compounds of the type *cis*-$[M(AA)_2X_2]^{n+}$ into optical isomers (Section 3–4). Another chemical test relies on the fact that typical bidentate ligands are able to span cis but not trans positions in a complex. The reaction of oxalate ion ($C_2O_4^{2-}$) with the cis and trans isomers of $[Pt(NH_3)_2Cl_2]$ illustrates this method. In Figure 4–2 it may be seen that the trans isomer forms a complex containing two oxalate ions each of which functions as a monodentate, whereas the cis isomer forms a complex that contains one bidentate oxalate. This approach has been most successful for platinum(II) complexes.

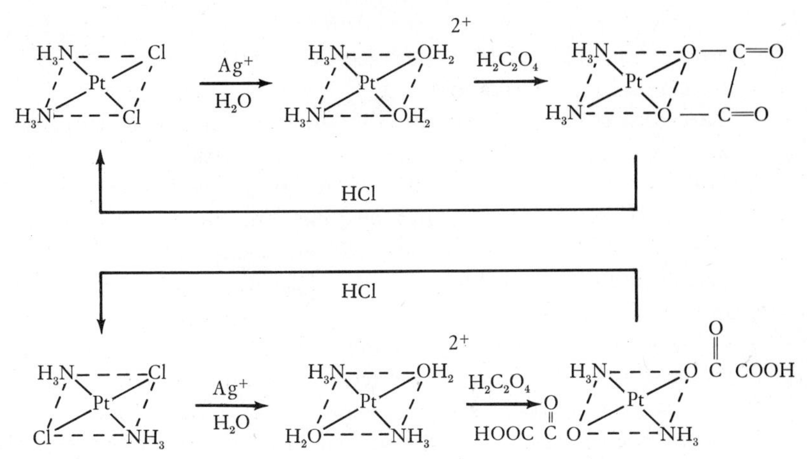

Figure 4–2 The reaction of oxalic acid with *cis*- and *trans*-[Pt(NH$_3$)$_2$Cl$_2$].

Structural assignments are now usually made on the basis of physicochemical techniques such as x-ray diffraction and spectroscopy. A relatively simple technique that has been used involves measurements of dipole moments. The dipole moments of cis and trans isomers are often markedly different. This seems to be particularly true for certain square planar complexes (48).

$(C_2H_5)_3P$ — C_6H_5 / Pt / $(C_2H_5)_3P$ — C_6H_5

$\mu = 7.2$
(in dipole units of Debye)

$(C_2H_5)_3P$ — C_6H_5 / Pt / C_6H_5 — $P(C_2H_5)_3$

$\mu \sim 0.0$

(48)

4–10 PREPARATION OF OPTICALLY ACTIVE COMPOUNDS

Many optically active organic molecules are present in plants and animals, and often they can be isolated and obtained pure. Laboratory preparations of compounds that can exhibit optical activity almost invariably yield 50–50 (racemic) mixtures of the two optical isomers and hence an optically inactive material (Section 3–4). Therefore, the basic step in the laboratory preparation of an optically active coordination compound is separation from its optical isomer. For example, racemic $[Co(en)_3]^{3+}$ is readily prepared by the air oxidation of a cobalt(II) salt in a medium containing excess ethylenediamine and a catalytic amount of activated charcoal. Since optical isomers are so very much alike, special separation techniques are required.

The most common techniques involve the principle that each of a pair of optical isomers will interact differently with a third optically active material. There is a subtle structural difference between optical isomers, and this will lead one isomer to be more strongly attracted to a third asymmetric molecule. For example, the salt *d*-$[Co(en)_3]$(*d*-tartrate)$Cl \cdot 5H_2O$ is less soluble than *l*-$[Co(en)_3]$(*d*-tartrate)$Cl \cdot 5H_2O$. This indicates that *d*-$[Co(en)_3]^{3+}$ forms a more stable crystalline lattice with *d*-tartrate than does *l*-$[Co(en)_3]^{3+}$. Therefore, the addition of a solution containing *d*-tartrate anion to a concentrated solution of racemic $[Co(en)_3]^{3+}$ causes the precipitation of *d*-$[Co(en)_3]$(*d*-tartrate)$Cl \cdot 5H_2O$. The *l*-$[Co(en)_3]^{3+}$ remains in solution and can be collected by the addition of I^-, which forms *l*-$[Co(en)_3]I_3$. This product will be contaminated with any *d*-$[Co(en)_3]^{3+}$ that was not removed in the tartrate precipitation.

The preferential precipitation of one of a pair of optical isomers by another optically active compound is the principal method for resolving pairs of optical isomers. The method can be used, however, only if the isomers to be separated can be obtained as charged ions, since their precipitation as salts is required.

It is difficult to resolve nonionic compounds, but compounds enriched in one optical isomer with respect to its enantiomer have been separated by a number of methods. All of the methods involve placing a material such as an optically active sugar or optically active quartz in a column (for example, a buret) and then passing a solution of the compound to be resolved (or a gaseous compound directly) through the column. It has been found that one isomer clings to the optically active packing in the column more tightly than the other. The less tightly bound isomer moves through the column more rapidly and hence comes off the column first. The complex [Cr(acac)$_3$] has been partially resolved by passing a solution of it in benzene-heptane (C_6H_6–C_7H_{16}) through a column packed with *d*-lactose, a naturally occurring sugar.

4–11 PREPARATION OF METAL CARBONYLS AND ORGANOMETALLIC COMPOUNDS

Compounds containing transition-metal–carbon bonds have been known for many years. The dye prussian blue ($Fe[Fe_2(CN)_6]_3$), which contains Fe—CN bonds, was perhaps the first example of such a coordination compound. The metal carbonyls $Ni(CO)_4$ and $Fe(CO)_5$ were prepared by the French chemist Mond in about 1890. Despite this early start, much of the progress in this field has been achieved since 1950. Since then compounds containing transition-metal carbon bonds have been synthesized in wide variety. Included among these compounds are alkyl [for example, $(CO)_5MnCH_3$] and aryl (for example, $[\{P(C_2H_5)_3\}_2Pt(C_6H_5)_2]$) compounds in which metal-carbon bonds (σ bonds) exist, a variety of olefin compounds (I) in which the bonding can be described as involving sharing

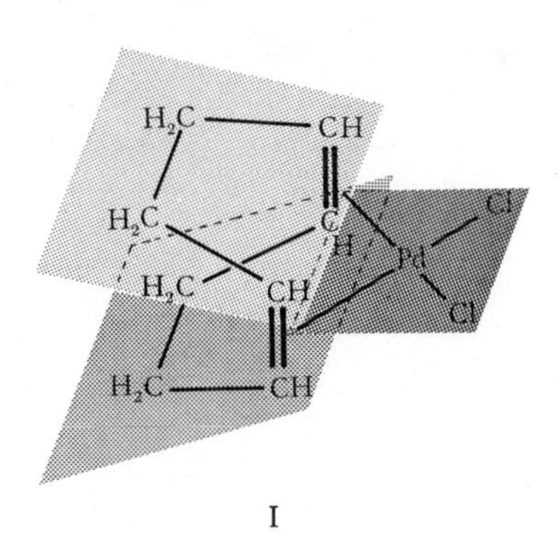

I

of π electrons of the olefin (an organic molecule containing double bonds) with the metal, and the so-called sandwich compounds (II) in which a metal is located between two planar (or nearly so) cyclic

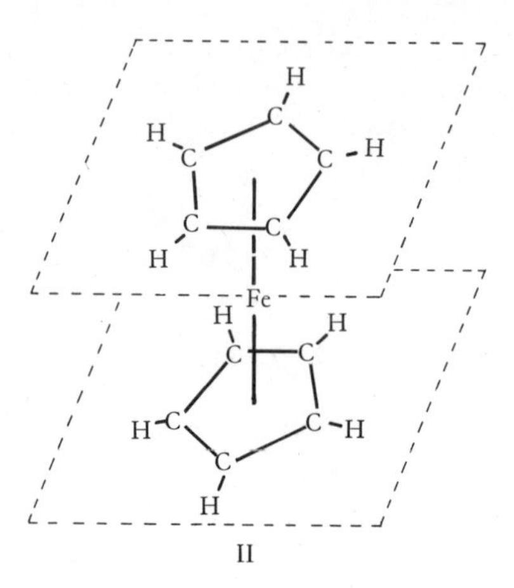

II

carbon compounds. In the latter compounds the bonding may be considered to involve the sharing of π electrons of the organic molecule with *d* or hybrid orbitals on the metal.

Metal atoms in compounds containing carbon-metal bonds frequently have formal oxidation states that are abnormally low; for example, the metals in the carbonyls are formally zero valent. This fact and their molecular nature suggest that the carbonyls might normally be prepared under reducing conditions, frequently in nonaqueous solvents. We shall find that many of these syntheses are performed in diglyme [$(CH_3OCH_2CH_2)_2O$], tetrahydrofuran ($\underline{CH_2CH_2CH_2CH_2O}$), or diethyl ether, solvents in which the reactants and products are soluble and which are more resistant to reduction than water.

Preparation of Metal Carbonyls

Metal carbonyls were first prepared by Mond by the direct reaction of gaseous carbon monoxide with finely divided metals. Iron, cobalt, and nickel carbonyls can be prepared in this way (49). The

$$\mathrm{Ni} + 4\mathrm{CO} \xrightarrow[25°\mathrm{C}]{1\ \mathrm{atm}} \underset{\text{colorless}}{\mathrm{Ni(CO)_4}} \quad (\mathrm{bp}\ 43°) \tag{49}$$

formation of $Ni(CO)_4$ proceeds rapidly at room temperature and 1 atm of CO. Higher pressures and temperatures are required to prepare $Co_2(CO)_8$. The Mond process for the metallurgy of nickel and cobalt separates these metals by the formation of $Ni(CO)_4$ at moderate temperatures and pressures and its subsequent decomposition to the metal and CO by stronger heating. Since $Co_2(CO)_8$ is formed only very slowly under the same conditions and has a relatively low volatility, cobalt is left behind when gaseous $Ni(CO)_4$ is removed.

A variety of metal carbonyls are known. Sidgwick's effective atomic number rule (Section 2–2) is very successful in explaining their stoichiometries. Simple monomeric carbonyls are expected for transition metals with even atomic numbers: $Cr(CO)_6$, $Fe(CO)_5$, $Ni(CO)_4$. The heavier members of the Cr and Fe families also form monomeric carbonyls of the predicted stoichiometry.

Transition metals with odd atomic numbers cannot achieve the expected EAN in monomeric compounds. In cases where carbonyls of these elements have been prepared, the compounds contain more than one metal atom and metal-metal bonds; these effectively contribute one extra electron to each metal: $Mn_2(CO)_{10}$, $Co_2(CO)_8$. Other polynuclear metal carbonyls are also known. Structures of some carbonyls are found in Figure 4–3. In 1959, $V(CO)_6$ was prepared. This compound is a black paramagnetic solid that decomposes at 70°. It is the only monomeric metal carbonyl that does not obey the EAN rule. The compound is easily reduced to $[V(CO)_6^-]$, which has 36 electrons around the metal (50).

$$\underset{\text{black}}{V(CO)_6} + Na \xrightarrow[\text{diglyme}]{} \underset{\text{yellow}}{[V(CO)_6]^-} + Na^+ \qquad (50)$$

Metal carbonyls are commonly made by reduction of metal salts in the presence of CO under high pressure. A variety of reducing agents have been found to be effective. In reaction (51) sodium reduction produces a solvated sodium salt of $V(CO)_6^-$; this is subsequently oxidized by H^+ to $V(CO)_6$ (52).

$$\underset{\text{pink}}{VCl_3} + 6Na + 6CO \xrightarrow[\substack{\text{diglyme} \\ \text{200 atm}}]{100^\circ} \underset{\text{yellow}}{[(\text{diglyme})_2Na]^+ [V(CO)_6]^-} \qquad (51)$$

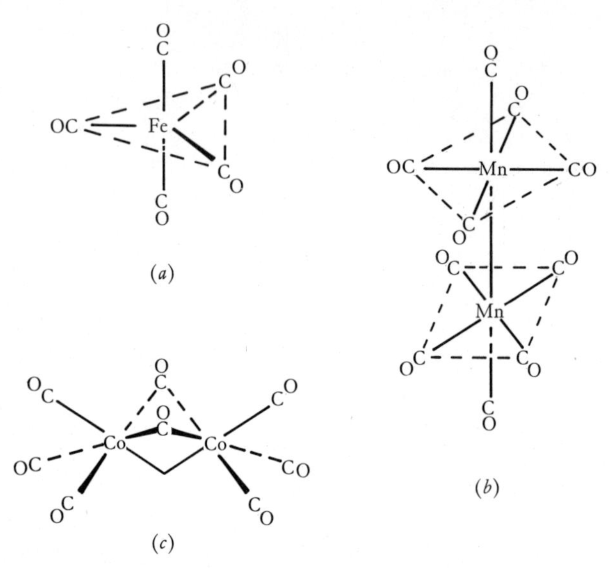

Figure 4–3 The structures of (*a*) $Fe(CO)_5$, (*b*) $Mn_2(CO)_{10}$, and (*c*) $Co_2(CO)_8$.

$$[(\text{diglyme})_2\text{Na}]^+[\text{V(CO)}_6]^- \xrightarrow[\text{H}_3\text{PO}_4(100\%)]{25^\circ} \text{V(CO)}_6 + \text{H}_2 \quad (52)$$

yellow (reactant); black ($V(CO)_6$)

Other active metals such as aluminum, as well as organic salts of active metals (for example, C_2H_5MgBr and C_6H_5Li), are often used. In a few cases metal carbonyls are conveniently prepared by reduction with the readily synthesized $Fe(CO)_5$ (53). Carbon

$$\text{WCl}_6 + 3\,\text{Fe(CO)}_5 \xrightarrow[\text{ether}]{90^\circ} \text{W(CO)}_6 + 3\text{FeCl}_2 + 9\text{CO} \quad (53)$$

blue (WCl_6); colorless ($W(CO)_6$)

monoxide itself is an excellent reducing agent and in some cases serves this function as well as that of a ligand (54).

$$\text{Re}_2\text{O}_7 + 17\text{CO} \xrightarrow[\text{and pressure of CO}]{\text{heat}} \text{Re}_2(\text{CO})_{10} + 7\text{CO}_2 \quad (54)$$

yellow-brown (Re_2O_7); colorless ($Re_2(CO)_{10}$)

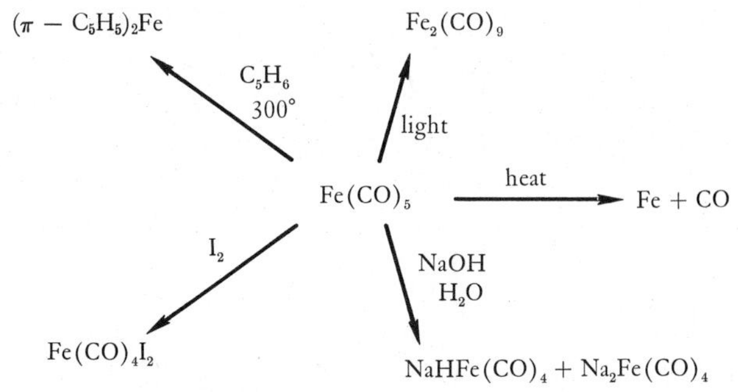

Figure 4–4 Reactions of $Fe(CO)_5$ illustrating reactions of metal carbonyls.

A variety of compounds can be obtained by reactions of the metal carbonyls. Reactions characteristic of these compounds are illustrated in Figure 4–4 by some reactions of $Fe(CO)_5$.

Preparation of Transition-Metal Olefin Compounds

In 1827 W. C. Zeise, a Danish pharmacist, found that ethylene, C_2H_4, reacts with $[PtCl_4]^{2-}$ in dilute HCl solution to yield compounds containing both platinum and ethylene. Only recently have the structures of the products been accurately determined, III and IV. These compounds can be described as being square planar with the platinum ethylene bonds pointing between the two carbon atoms.

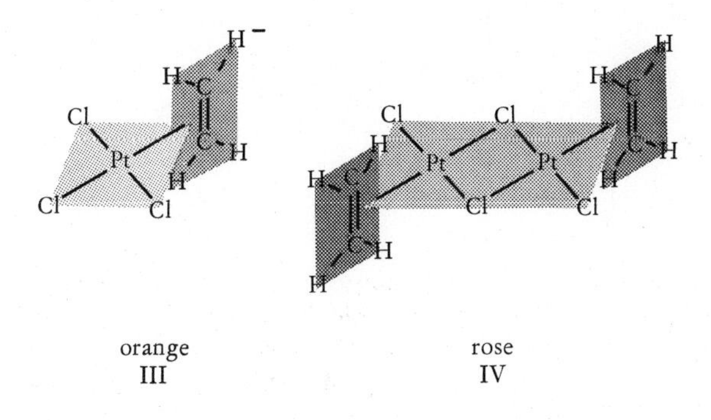

Molecular orbital theoretical treatments of ethylene (and other olefin) metal complexes describe the bonding as involving an overlap of an empty metal orbital with a filled π molecular orbital that is delocalized over the ethylene molecule. Additional stability is provided by possible π bonding between suitably oriented filled metal orbitals and vacant antibonding olefin molecular orbitals (Figure 4–5).

In recent years a wide variety of olefin and related compounds have been prepared. The most stable of these seem to be formed by molecules containing two double bonds that are so located that they can both bond to the same metal. One molecule of this type is cyclooctadiene (55). Olefin compounds are normally prepared by the direct reaction of an olefin with a metal salt or complex.

Preparation of Sandwich Compounds

Since 1950, a large number of transition-metal compounds have been prepared in which the metal atom has been described as the "meat" between two flat organic molecules that are the "slices of bread" in the "sandwich" molecule. A variety of metals and or-

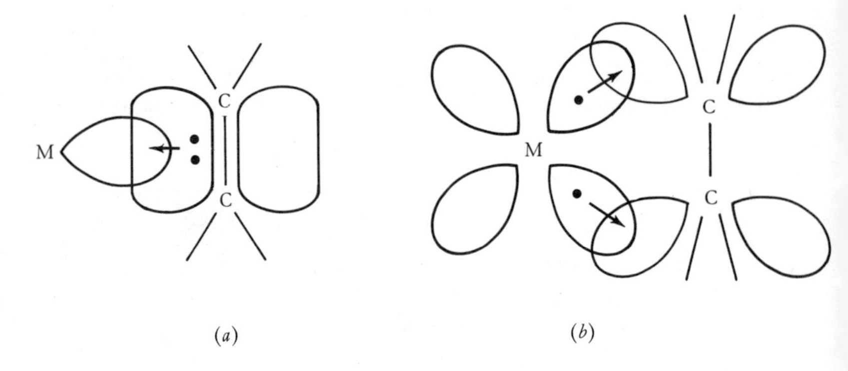

Figure 4–5 A representation of the bonding in metal olefin complexes: (*a*) the σ bond in which a π MO on the olefin overlaps with a metal orbital; (*b*) the π bond in which the π^* antibonding MO of the olefin overlaps with a metal *d* orbital.

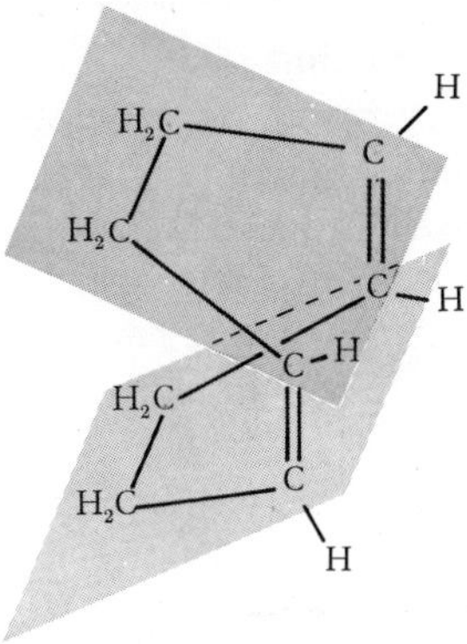

$$+ \ RhCl_3 \cdot 3H_2O \xrightarrow[85°]{C_2H_5OH}$$

red

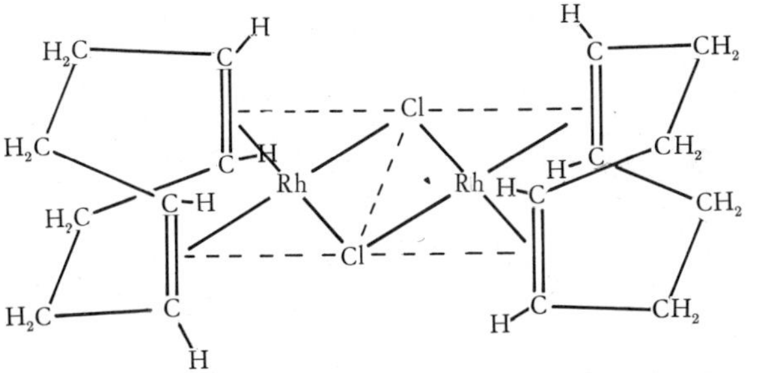

(55)

orange

ganic molecules have been used successfully; the most stable compounds contain the cyclopentadiene anion, $C_5H_5^-$, V.

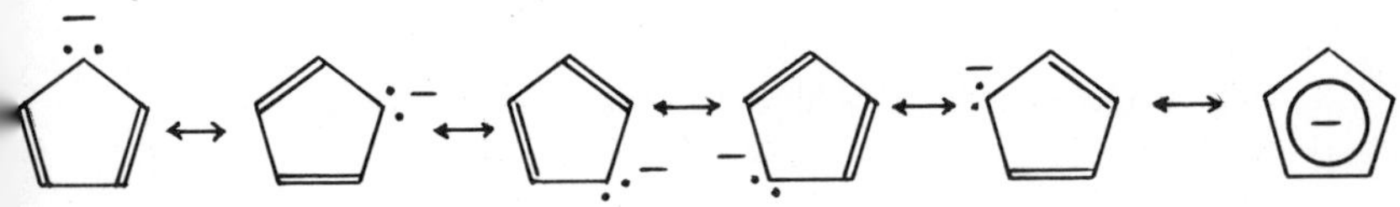

V

The first compound recognized to be of this type was ferrocene [bis(π-cyclopentadienyl)iron(II)], II. It is an orange crystalline compound that boils without decomposition at 249°C and is unaffected by aqueous NaOH or concentrated HCl. The compound is diamagnetic and nonpolar. On the basis of such chemical and physical properties, the sandwich structure II was proposed. Structure II has been confirmed by x-ray diffraction studies.

Sidgwick's EAN rule has been very useful in the preparation of sandwich and olefin compounds. The $C_5H_5^-$ ion is treated as a six-electron donor, as is the benzene molecule; ethylene is a two-electron donor. Combinations of ligands that donate the proper number of electrons to a metal to give it an EAN equal to that of a rare gas frequently produce stable compounds, for example, $Fe(C_5H_5)_2$, $Mn(C_5H_5)(C_6H_6)$, $Cr(C_6H_6)_2$.

In the ferrocene molecule the cyclopentadiene anion reacts like an aromatic organic molecule. Since ferrocene is quite stable, it has been possible to perform the reactions characteristic of an aromatic system on the ring without destroying the bonding to the metal (56). An extensive organic chemistry of ferrocene, with a corresponding production of new ferrocene derivatives, has been developed.

$$\underset{\text{orange}}{\mathrm{Fe}(C_5H_5)_2} + CH_3COCl \xrightarrow[CS_2]{AlCl_3} \underset{\text{orange}}{\mathrm{Fe}(C_5H_5)(C_5H_4\text{–}\overset{\overset{\displaystyle O}{\|}}{C}\text{–}CH_3)} \qquad (56)$$

The cyclopentadiene sandwich compounds are often prepared by the reactions of metal halides with sodium cyclopentadienide (57).

$$\underset{\text{green}}{FeCl_2} + C_5H_5Na \xrightarrow{\text{ether}} \underset{\text{orange}}{(C_5H_5)_2Fe} + 2NaCl \qquad (57)$$

Dibenzenechromium, $Cr(C_6H_6)_2$, is best prepared by the reaction scheme (58). An intermediate Cr(I) compound is produced; it is then treated with a strong reducing agent such as dithionite ion

$$3CrCl_3 + 2Al + AlCl_3 + 6C_6H_6 \rightarrow 3[(C_6H_6)_2Cr]^+[AlCl_4]^-$$
violet — yellow

$$(C_6H_6)_2Cr + SO_3^{=} \xleftarrow[OH^-]{S_2O_4^{2-}} [(C_6H_6)_2Cr]^+ClO_4^- \quad (58)$$
($ClO_4^- \downarrow$ from the yellow tetrachloroaluminate)
brown — yellow

$(S_2O_4^{2-})$. Dibenzene chromium and other benzene sandwich compounds are considerably less stable than many of the cyclopentadiene compounds. They are much more readily oxidized and decomposed under most reaction conditions.

Preparation of Transition-Metal–Carbon σ-Bonded Compounds

Transition-metal compounds containing σ-bonded alkyl or aryl groups were rare until recently. It has been found that the presence of ligands such as CO, $C_5H_5^-$, or phosphines on transition metals greatly enhances the ability of the transition metals to form σ-bonded organometallic compounds. Transition-metal–carbon σ bonds are often produced by metathesis reactions in which one product is the organometallic compound and the other is a simple salt (59), (60).

$$(CO)_5MnNa + CH_3I \rightarrow (CO)_5MnCH_3 + NaI \quad (59)$$
colorless — colorless

$$cis\text{-}[\{P(C_2H_5)_3\}_2PtBr_2] + 2CH_3MgBr \xrightarrow{\text{ether}} \quad (60)$$
colorless

$$cis\text{-}[\{P(C_2H_5)_3\}_2Pt(CH_3)_2] + 2MgBr_2$$
colorless

Transition-metal fluorocarbon derivatives have been prepared recently; in general, they are more stable than the corresponding hydrocarbon compounds (61), (62).

$$HMn(CO)_5 + C_2F_4 \xrightarrow[25^\circ]{\text{pentane}} HCF_2CF_2Mn(CO)_5 \quad (61)$$
colorless — colorless

$$Fe(CO)_5 + F_3CCF_2I \xrightarrow[45^\circ]{\text{benzene}} F_3CCF_2Fe(CO)_4I \quad (62)$$
yellow — purple

The chemistry of transition-metal organometallic compounds has been only briefly touched upon here, but there is a great deal of current research interest in the field. The chemistry of these compounds is of considerable practical importance, since the reactions of the compounds are believed to play an important role in the activity of transition-metal catalysts in a variety of organic systems.

PROBLEMS

1. Write appropriate balanced equations and give approximate experimental conditions for the preparation of each of the following:

$[Ir(NH_3)_5ONO]^{2+}$ starting with $[Ir(NH_3)_5OH_2]^{3+}$
$[Pt(NH_3)_4]Cl_2$ starting with K_2PtCl_4
$[Co(NH_3)_5Br]Br_2$ starting with $CoBr_2 \cdot 6H_2O$
$[Co(en)_3]Cl_3$ starting with $[Co(NH_3)_5Cl]Cl_2$
trans-$[Pt(NH_3)_4(OH)_2]SO_4$ starting with $[Pt(NH_3)_4]SO_4$
$[Cu(gly)_2]$ starting with $CuCO_3$
$[Ni(DMG)_2]$ starting with $NiCl_2 \cdot 6H_2O$
$Ru(C_5H_5)_2$ starting with $RuCl_2$
$Mn(CO)_5Br$ starting with $Mn_2(CO)_{10}$
$K[Pt(C_2H_4)Cl_3]$ starting with K_2PtCl_4

2. Complete and balance the following equations. Unless otherwise stated, the reactions proceed at room temperature in water solution.

$AgCl + NH_3 \rightarrow$
$[Ni(NH_3)_6]^{2+} + CN^- \rightarrow$
$[Co(H_2O)_6]^{2+} + NH_3 + NH_4Cl + PbO_2 \rightarrow$
$[Rh(NH_3)_5H_2O]^{3+} + OH^- \xrightarrow[\text{cold water}]{\text{fast}}$
$[Co(NH_3)_5Cl]^{2+} + [Ag(H_2O)_2]^+ \xrightarrow{\text{heat}}$
$[Pt(PCl_3)_2Cl_2] + C_2H_5OH \rightarrow$ (C_2H_5OH solvent)
$Ni(CO)_4 + O_2 \xrightarrow{\text{heat}}$ (gaseous reaction)
$Fe(CO)_5 + Br_2 \rightarrow$ (no solvent)
$Cr(CO)_6 \xrightarrow{\text{heat}}$ (no solvent)

3. Identify the colored species designated here by capital letters. Excess ammonia was added to a pink aqueous solution (A) in the absence of air to give a pale-blue solution (B). In the presence of air the color of this solution slowly changed and eventually became rose (C). The addition of charcoal to the boiling solution (C) resulted in a yellow solution (D). Addition of excess HCl to this yellow solution caused the precipitation of a yellow-orange crystalline product (E).

4. (*a*) By making use of the trans-effect phenomenon in platinum(II) complexes, show how the following complexes may be prepared starting with K_2PtCl_4. (*b*) Show the method of preparation

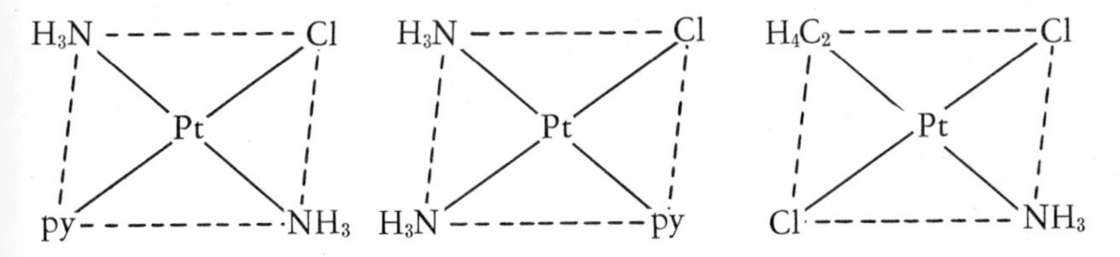

for all of the possible isomers of $[Pt(py)(NH_3)(I)(Cl)]$.

REFERENCES

The preparations of inorganic compounds have been published in separate volumes of *Inorganic Syntheses* starting in 1939. The seven volumes now available include many examples of metal complexes. Detailed specific directions are given.

Inorganic Syntheses, McGraw-Hill, New York, 7 vols., 1939 to 1963.

H. F. Walton, *Inorganic Preparations*, Prentice-Hall, Englewood Cliffs, N.J., 1948.

W. G. Palmer, *Experimental Inorganic Chemistry*, Cambridge, New York, 1954.

W. L. Jolly, *Synthetic Inorganic Chemistry*, Prentice-Hall, Englewood Cliffs, N.J., 1960.

G. G. Schlessinger, *Inorganic Laboratory Preparations*, Chemical Publishing, New York, 1962.

V

Complex Ion Stability

To understand the solution chemistry of metals, we must know the nature and the stability of the complexes that metal ions can form with the solvent and with potential ligands in solution. Research in this area provides the data needed to obtain a clearer insight into the factors that contribute to the stability of metal complexes. Many important applications of this information are made. The reader will recall that in qualitative analysis certain precipitates are dissolved by using an appropriate complexing agent. The use of hypo, $Na_2S_2O_3$, in the fixing process in the photographic industry is effective because the silver halide in the film emulsion is dissolved owing to the formation of the stable and soluble silver thiosulfate complex $[Ag(S_2O_3)_2]^{3-}$. The addition of complexing agents to hard water results in the generation of stable and soluble complexes of the objectionable metal ions, such as calcium, which in turn prevents the precipitation of the insoluble metal salts of common soaps (soap scum).

When ammonia is added to a solution of a copper(II) salt, there is a rapid reaction in which water coordinated to the metal ion is replaced by ammonia. Although the product of this reaction is normally represented as $[Cu(NH_3)_4]^{2+}$, in fact a variety of products result, the relative amount of each species depending upon the concentrations of cupric ion and ammonia (1) to (4). Figure 5–1 is a plot of the percentage of each copper(II) ammonia species in solution vs.

$$[Cu(H_2O)_4]^{2+} + NH_3 \rightleftharpoons [CuNH_3(H_2O)_3]^{2+} \quad (1)$$

$$[CuNH_3(H_2O)_3]^{2+} + NH_3 \rightleftharpoons [Cu(NH_3)_2(H_2O)_2]^{2+} \quad (2)$$

$$[Cu(NH_3)_2(H_2O)_2]^{2+} + NH_3 \rightleftharpoons [Cu(NH_3)_3(H_2O)]^{2+} \quad (3)$$

$$[Cu(NH_3)_3H_2O]^{2+} + NH_3 \rightleftharpoons [Cu(NH_3)_4]^{2+} \quad (4)$$

the concentration of free ammonia. This plot indicates that $[Cu(NH_3)_4]^{2+}$ is a very important species, since it is the predominant one in solutions containing 0.01 to 5 *M* free ammonia. Outside this range, however, the other ammine complexes are more abundant.

On a strictly statistical basis one would expect that the relative number of H_2O and NH_3 molecules surrounding Cu^{2+} would be the same as the relative number of NH_3 and H_2O molecules in

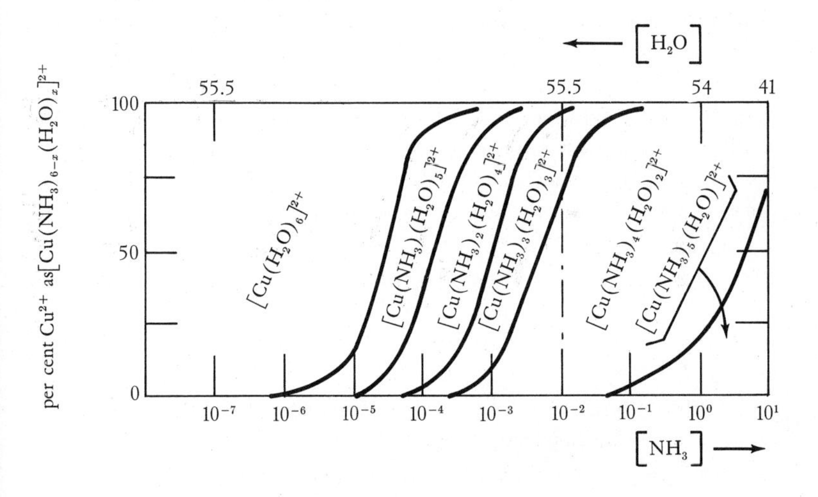

Figure 5–1 A plot of the percentage of Cu^{2+} present in the form of various ammine complexes for solutions containing different concentrations of free ammonia. (For example, at a free NH_3 concentration of 1.0×10^{-3} *M*, the Cu^{2+} is present as 65 per cent $[Cu(NH_3)_4(H_2O)_2]^{2+}$, 30 per cent $[Cu(NH_3)_3(H_2O)_3]^{2+}$, and 5 per cent $[Cu(NH_3)_2(H_2O)_4]^{2+}$).

solution; that is, if the solution contains equal numbers of NH_3 and H_2O molecules, the predominant cupric species should be $[Cu(NH_3)_2(H_2O)_2]^{2+}$. This type of statistical ligand distribution is not observed. Metal ions exhibit marked preferences for certain ligands; for example, cupric ion coordinates NH_3 in preference to H_2O. Nonetheless, statistical considerations are also important in that an increase in NH_3 concentration produces copper(II) species containing a larger number of coordinated NH_3 molecules.

In some cases the preferences of metals for certain ligands is readily understood; it seems reasonable that positive metal ions should prefer anionic ligands to neutral or positively charged ones. In general, however, the factors that determine which ligand will coordinate best with a given metal ion are numerous and complicated and are not completely understood. Later in this chapter some of these factors will be discussed.

The properties of a metal ion in solution are dependent on the nature of the groups (ligands) surrounding the metal. The number and type of such groups cannot be predicted on a statistical basis. Consequently, many studies have been made to establish the composition of the coordination spheres of metal ions in solution in the presence of a wide variety of possible ligands. The information collected in these studies is conveniently expressed by *stability constants*. These constants are closely related to the familiar expressions used to describe the ionization of acids and bases (5), (6).[1] In fact

$$HF \rightleftharpoons H^+ + F^- \tag{5}$$

$$K_{ionization} = \frac{[H^+][F^-]}{[HF]} \tag{6}$$

a complex is broadly defined as a species formed by the association of two or more simpler species each capable of independent existence. In this sense an acid is a complex of the "metal" H^+ and another species, frequently an anion. An *ionization constant* is a type of *stability constant* (actually an *instability constant*). The formulation and application of stability constants is presented in Section 5–1.

[1] Brackets are used in (6) to designate concentration. Thus the symbol $[A^-]$ indicates the concentration of A^- in moles per liter.

In discussions of the solution behavior of coordination compounds it is normally assumed that the solvent is water, but nonaqueous solvents will dissolve certain coordination compounds and are being more widely used. In these solvents solute species are surrounded by solvent molecules, and complexation reactions then involve the replacement of solvent molecules by other ligands. In principle, equilibrium behavior in nonaqueous solvents can be handled in a manner analagous to that of water. The limited solubility of ionic species in most nonaqueous solvents, problems associated with the lack of dissociation of salts (ion pairing) in most nonaqueous solvents, and the convenience of aqueous systems have resulted in the use of aqueous media in most equilibrium studies. In the following discussion aqueous equilibria will be specifically considered, but with certain modifications the same treatment will apply in other solvents.

5–1 STABILITY CONSTANTS

It has been found that in a reaction mixture at equilibrium at a certain temperature the product of the activities of the products di-

$$a\text{A} + b\text{B} + \cdots \rightleftharpoons c\text{C} + d\text{D} + \cdots$$

vided by the produ ct of the activities of the reactants is equal to a constant, Equation (7), called γ the *equilibrium constant* for the given

$$K = \frac{a_{\text{C}}{}^{c} a_{\text{D}}{}^{d} \cdots}{a_{\text{A}}{}^{a} a_{\text{B}}{}^{b} \cdots} = \text{constant} \tag{7}$$

reaction. The *activity* of a species A is the product of its concentration and an *activity coefficient* γ_A.

$$a_{\text{A}} = [\text{A}]\gamma_{\text{A}}$$

The activity coefficient has a value of unity in very dilute solutions, so that under such conditions concentrations and activities are numerically equal. In the 0.01 to 5 M solutions most commonly used in the laboratory, activity coefficients are less than 1, and hence activities are lower than concentrations.

That the activity of a species in solution is less than its concentration is interpreted as indicating that the species cannot act independently while it is under the influence of other solute particles. Hence its effective concentration is decreased. In subsequent discussions of equilibrium constants we shall frequently replace activities with concentrations. One should keep in mind that this involves the assumption of unit activity coefficients, and thus will be quantitatively accurate only in very dilute solution.

An equilibrium constant is very useful in that it summarizes a great deal of information. For example, the equilibrium constant for reaction (8) was measured in a 0.128 *M* $HClO_4$ solution and found

$$[Fe(H_2O)_6]^{3+} + NCS^- \rightleftharpoons [Fe(H_2O)_5NCS]^{2+} + H_2O \qquad (8)$$

to be 234. From this number one can calculate the concentration of Fe^{3+}, NCS^-, or $Fe(NCS)^{2+}$ in a solution when the concentrations of the other two species are known.[1] A more useful problem is the calculation of the concentrations of all three species in a solution made up from a solution of 0.0100 *M* Fe^{3+} in 0.128 *M* $HClO_4$ to which enough KNCS had been added to bring the total concentration of free and complexed thiocyanate ion to 0.0100 *M*.[2]

$$[Fe^{3+}] + [FeNCS^{2+}] = 0.0100\ M$$

$$[NCS^-] + [FeNCS^{2+}] = 0.0100\ M$$

$$[Fe^{3+}] + [FeNCS^{2+}] = [NCS^-] + [FeNCS^{2+}] = 0.0100\ M$$

$$\therefore [Fe^{3+}] = [NCS^-]$$

$$234 = \frac{[Fe(NCS)^{2+}]}{[Fe^{3+}][NCS^-]} = \frac{[0.0100 - [Fe^{3+}]]}{[Fe^{3+}]^2}$$

$$234[Fe^{3+}]^2 + [Fe^{3+}] - 0.0100 = 0$$

$$[Fe^{3+}] = \frac{-1.00 + \sqrt{(1.00)^2 - 4(234)(-0.0100)}}{2(234)}$$

$$= 0.0047\ M$$

$$[NCS^-] = 0.0047\ M$$

$$[FeNCS^{2+}] = 0.0100 - 0.0047 = 0.0053\ M.$$

[1] Fe^{3+} and $FeNCS^{2+}$ are used in place of $[Fe(H_2O)_6]^{3+}$ and $[Fe(NCS)(H_2O)_5]^{2+}$. One should realize that all ions are hydrated in aqueous solution.

[2] The perchloric acid ($HClO_4$) is present to prevent the acid dissociation of the hydrated Fe^{3+} ion (9).

$$[Fe(H_2O)_6]^{3+} \rightleftharpoons [Fe(H_2O)_5OH]^{2+} + H^+ \qquad (9)$$

This calculation gives only an approximate answer, since the formation of species such as $[Fe(NCS)_2]^+$ was neglected. Knowledge of necessary equilibrium constants for solutions of metal ions and potential ligands enables one to calculate the concentrations of all species in solution.

Metal complexes are formed in solution by stepwise reaction, and equilibrium constants can be written for each step (10), (11).

$$Ag^+ + NH_3 \rightleftharpoons Ag(NH_3)^+ \qquad K_1 \tag{10}$$

$$Ag(NH_3)^+ + NH_3 \rightleftharpoons [Ag(NH_3)_2]^+ \qquad K_2 \tag{11}$$

The water molecules that make up the hydration sphere of an aqueous metal ion are often omitted in writing the reaction equation, since frequently the number of coordinated water molecules is unknown. Moreover, the water molecules involved in the reaction are never included in the equilibrium constant. Since the activity of pure water is by convention defined as unity (although its concentration is 55.6 M), it follows that in dilute solutions the activity of water is close to 1.

The equilibrium constants K_1 (12), and K_2 (13) are called *step-*

$$K_1 = \frac{[Ag(NH_3)^+]}{[Ag^+][NH_3]} \tag{12}$$

$$K_2 = \frac{[Ag(NH_3)_2{}^+]}{[Ag(NH_3)^+][NH_3]} \tag{13}$$

wise stability constants. They are called stability constants because the larger the value of the constant, the greater the concentration of the complex species at equilibrium. An acid dissociation constant is an *instability constant* (14), because it describes the dissociation of

$$HX \rightleftharpoons H^+ + X^- \qquad K_a = \frac{[H^+][X^-]}{[HX]} \tag{14}$$

the acid (complex). The exact opposite is true for a stability constant which is a measure of extent of association. Stability constants are normally used to describe the equilibrium behavior of metal complexes.

A second type of equilibrium constant β, called an *over-all stability constant*, is also used (15). Since K's and β's describe exactly the

$$\beta_1 = \frac{[Ag(NH_3)^+]}{[Ag^+][NH_3]} \qquad \beta_2 = \frac{[Ag(NH_3)_2{}^+]}{[Ag^+][NH_3]^2} \tag{15}$$

same chemical systems, they must be related to each other (16).

$$\begin{aligned} \beta_2 &= \frac{[Ag(NH_3)_2{}^+]}{[Ag^+][NH_3]^2} = \frac{[Ag(NH_3)_2{}^+]}{[Ag^+][NH_3][NH_3]} \cdot \frac{[Ag(NH_3)^+]}{[Ag(NH_3)^+]} \\ &= \frac{[Ag(NH_3)_2{}^+]}{[Ag(NH_3)^+][NH_3]} \cdot \frac{[Ag(NH_3)^+]}{[Ag^+][NH_3]} = K_2K_1 \end{aligned} \tag{16}$$

From (15) it is seen that $\beta_1 = K_1$, and from (16) that $\beta_2 = K_1K_2$. In general $\beta_n = K_1K_2 \cdots K_n$.

The numerical value of a stability constant describes the relative concentrations of species at equilibrium. Large stability constants indicate that the concentration of complex is much greater than the concentrations of the components of which it is made. We say that a complex is stable if the equilibrium constant describing its formation is large. We shall see in Chapter VI that this need not imply that the compound is slow to react or that the ligands are resistant to replacement by ligands other than water.

5-2 FACTORS THAT INFLUENCE COMPLEX STABILITY

The term "stable complex" is defined in terms of the equilibrium constant for the formation of the complex. In the language of thermodynamics the *equilibrium constant* of a reaction is *the measure* of the *heat released* in the reaction and the *entropy change* during the reaction. The greater the amount of heat evolved in a reaction the more stable are the reaction products. The *entropy* of a system is a measure of the amount of disorder. The greater the amount of disorder in the products of a reaction relative to the reactants, the greater the increase in entropy during the reaction and the greater

TABLE 5–1

Factors that Increase the Stability of a Complex $[ML_x]^{n+}$ and Factors that Increase the Value of β

$$M^{n+} + xL \rightleftharpoons [ML_x]^{n+} + \Delta$$

$$\beta_x = \frac{[ML_x{}^{n+}]}{[M^{n+}][L]^x}$$

(*a*) An increase in the heat Δ evolved in the reaction.

(*b*) An increase in the amount of disorder produced during the reaction—an increase in entropy as the reaction proceeds.

---

the stability of the products. Table 5–1 summarizes the effect of the heat of reaction and changes in entropy on the stability of reaction products. Now let us consider separately the influence of the heats and entropies of complexation reactions on the stability of the complex formed.

The relative stabilities of many complexes can be understood in terms of a simple electrostatic model. The predictions of this model are related primarily to the heat evolved in the formation of the complexes. We are familiar with the observation that oppositely charged particles attract one another, whereas like charged particles repel one another. Moreover, the repulsion or attraction depends upon the distance between the centers of the particles, being greater as the particles approach one another.

One would therefore predict that the most stable complexes would be made up of oppositely charged ions and, moreover, that the greater the charge on the ions and the smaller the ions the greater should be the stability. Small ions are favored because their centers can be closer together. From this point of view the stability of complexes should increase with the charge on the metal ions. One illustration of this behavior is the increase of stability of hydroxide complexes with an increase in charge of the metal ion.[1] The stability

[1] Stability constants reported in this book are values at 25°C and are activity constants unless otherwise specified. Since literature values are often not in agreement, generally only one significant figure will be used.

$K_{LiOH} = 2 \qquad K_{MgOH^+} = 10^2 \qquad K_{YOH^{2+}} = 10^7 \qquad K_{ThOH^{3+}} = 10^{10}$

of complexes of metal ions having the same charge should increase as the ionic radius decreases. The stability constants of MOH^+ for the alkaline-earth metals illustrate this trend. From these data one can

$K_{BeOH^+} = 10^7 \qquad K_{MgOH^+} = 120 \qquad K_{CaOH^+} = 30 \qquad K_{BaOH^+} = 4$

see that the stability constants of YOH^{2+} and $BeOH^+$ are about the same. Thus a very small doubly charged cation can form complexes with stability comparable to that of complexes of larger, more highly charged cations.

A convenient criterion for the estimation of the complexing ability of metal ions is the charge-to-radius ratio. Table 5–2 shows the correlation between charge-to-radius ratio and the stability of OH^- complexes. From Table 5–2 it appears that the charge on the metal ion is somewhat more important than the ionic radius, but an adequate prediction of stability can often be made in this way.

The stabilities of high-spin complexes of the M^{2+} ions between

TABLE 5–2

The Effect of Cation Charge and Radius on the Stability of the Hydroxide ($MOH^{(n-1)+}$) Complex of the Cation

$M^{n+} + OH^- \rightleftharpoons MOH^{(n-1)+}$

M^{n+}	*Ionic radius*	$\frac{\textit{Charge}}{\textit{radius}}$	$K_{MOH^{(n-1)+}}$
Li^+	0.60	1.7	2
Ca^{2+}	0.99	2.0	3×10^1
Ni^{2+}	0.69	2.9	3×10^3
Y^{3+}	0.93	3.2	1×10^7
Th^{4+}	1.02	4.0	1×10^{10}
Al^{3+}	0.50	6.0	1×10^9
Be^{2+}	0.31	6.5	1×10^7

Mn^{2+} and Zn^{2+} with a given ligand frequently vary in the order $Mn^{2+} < Fe^{2+} < Co^{2+} < Ni^{2+} < Cu^{2+} > Zn^{2+}$ (see Figure 5–2). This order, sometimes called the *natural order of stability*, is relatively consistent with the charge-to-radius concept, since the radii of the ions vary in the order $Mn^{2+} > Fe^{2+} > Co^{2+} > Ni^{2+} < Cu^{2+} < Zn^{2+}$. Both the variation in size of the cation and the order of stability can be explained in terms of the crystal field stabilization energy (CFSE) for these complexes (Section 2–5). The high-spin complexes of these six metals are primarily octahedral with the exception of those of Cu^{2+} which, as we have noted, form tetragonally distorted octahedra. In an octahedral crystal field, electrons in the three t_{2g} *d* orbitals have a lower energy than electrons in the two e_g *d* orbitals (Figure 2–8). The t_{2g} orbitals are $0.4\Delta_o$ lower in energy than the hypothetical five degenerate *d* orbitals that were postulated prior to CF splitting; the e_g orbitals are $0.6\Delta_o$ higher. When one places an electron in a t_{2g} orbital rather than in one of the five degenerate *d* orbitals, one obtains stabilization of the magnitude $0.4\Delta_o$. It may be said that the

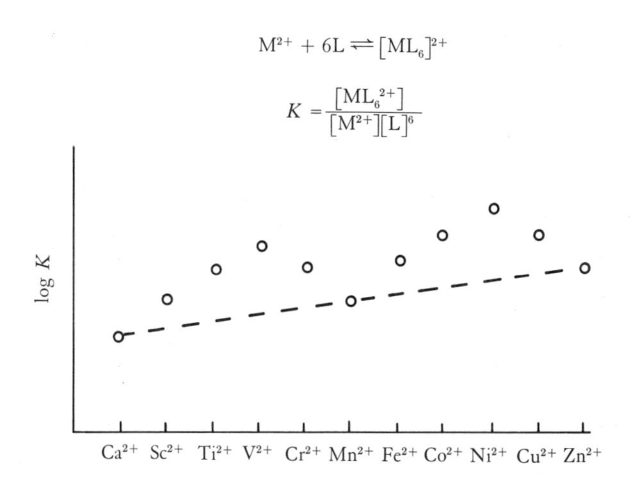

Figure 5–2 The logarithms of stability constants for a series of $[ML_6]^{2+}$ complexes as predicted by crystal field theory. Without the concepts of CF theory the values would be expected to increase fairly regularly along the dashed line. Experimental results show the double maxima represented by the circles.

system has saved $0.4\Delta_o$ in energy owing to CF splitting, or that the CFSE for this system is $0.4\Delta_o$. The CFSE's for octahedral complexes containing larger numbers of d electrons are indicated in Table 2–1.

Figure 5–2 indicates the relative stability of high-spin octahedral $[M(II)L_6]$ complexes of the first-row transition elements as predicted by CFT. The d^3 and d^8 systems will be the most stable with respect to their neighbors, since they have the greatest CFSE. As one progresses from Ca^{2+} to Zn^{2+} complexes, a general increase in stability is observed; this results from the decrease in the radii of the M^{2+} ions as one progresses toward Zn^{2+}. The order of stability predicted by CFT and presented in Figure 5–2 parallels the natural order of stability for complexes of these metals except for Cu^{2+}, and the natural order of stability can thus be attributed to CF stabilization. The discrepancy for Cu^{2+} is not completely understood, but it is certainly related to the fact that Cu^{2+} complexes assume a distorted octahedral structure in order to achieve maximum CFSE.

The electrostatic influence of the charge and size of the ligand are also important in determining the stability of a complex. For example, the small fluoride ion forms more stable Fe^{3+} complexes than does the larger chloride ion. However, most ligands consist of

$$K_{FeF^{2+}} = 1 \times 10^6 \qquad K_{FeCl^{2+}} = 2 \times 10^1$$

several atoms; therefore, it is difficult to assign a meaningful ligand radius and hence apply this criterion of size. It is interesting to note that the large singly charged perchlorate ion ClO_4^- has a very small tendency to form metal complexes, which is in keeping with the electrostatic point of view.

A number of important ligands are neutral molecules (H_2O, NH_3, H_2S, etc.). In terms of electrostatics these ligands are bound to metal ions through the attraction between the negative end of the ligand dipole and the metal cation (I). The more polar the ligand, the greater should be the force binding the ligand and metal ion. Water is the most polar of the common ligands and hence would be expected to form metal complexes of greater stability than other neutral ligands. The fact that water is the best solvent for many salts is in part a result of the stable complexes that it forms with metal ions.

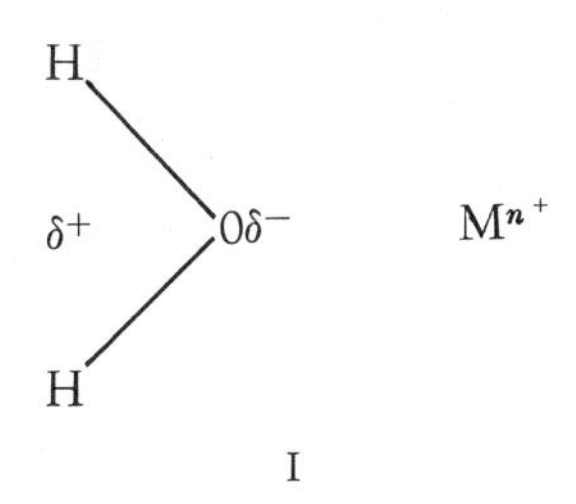

I

It has been observed that the greater the base strength of a ligand, the greater is the tendency of the ligand to form stable metal complexes. The base strength of a molecule is a measure of the stability of the "complex" that the molecule forms with H^+. It is reasonable that ligands that bind H^+ firmly should also form stable complexes with metal ions. From this point of view F^- should form more stable complexes than Cl^-, Br^-, or I^-, and NH_3 should be a better ligand than H_2O, which in turn should be better than HF. This predicted behavior is observed for the alkali and alkaline-earth metals and for other electropositive metals such as the first-row transition elements and the lanthanide and actinide elements. These metals are often called *class a metals*.

The simple electrostatic approach that has been outlined is successful in explaining the observed stability of many metal complexes and also in predicting the stabilities of other compounds. It is particularly effective for complexes of class *a* metal ions. In complexes of class *b* metal ions—ions of the more electronegative elements such as Pt, Au, Hg, and Pb and some lighter transition metals in low oxidation states—the electrostatic contributions are still important but other factors also play a role. In particular, crystal field effects and covalent bonding are also important. For example, Co^{2+}, Ni^{2+}, and Cu^{2+} prefer NH_3 to H_2O as a ligand. This is due at least partially to the fact the NH_3 provides a greater crystal field than H_2O (Section 2–5). Likewise we find that certain transition elements form very stable complexes with ligands such as CO, CN^-, C_2H_4, and $P(CH_3)_3$, whereas these are very poor ligands for non-transition elements. The stability of these transition-metal complexes can again be attributed to the crystal field stabilization provided by the ligands.

Metal-ligand covalent bonding becomes increasingly important in complexes of relatively electronegative metals such as those in the copper and zinc families and tin and lead. For these metals an electrostatic approach to stability is often ineffective. For example, silver forms insoluble halide salts, AgX, and stable halide complexes, AgX_2^- and AgX_3^-, in which the order of stability is $I^- > Br^- > Cl^- >> F^-$. The stability constants for reaction (17) are

$$K_{AgF} = 2 \qquad K_{AgCl} = 2 \times 10^3 \qquad K_{AgBr} = 3 \times 10^4 \qquad K_{AgI} = 10^8.$$

$$Ag^+ + X^- \rightleftharpoons AgX \text{ (aqueous)} \tag{17}$$

This behavior is attributed to covalent character in the Ag—X bond, which increases as one goes from F^- to I^-.

Mercury, lead, bismuth and other transition and post-transition metals form water-insoluble sulfide salts. The precipitation of these sulfides is part of the traditional method for the qualitative determination of metals. The formation of a precipitate can be considered as the formation of a zero-charged, water-insoluble complex; the formation of the sulfide precipitates indicates that these metals prefer sulfur-containing ligands to oxygen-containing ligands (in this case S^{2-} to O^{2-}). This preference for sulfur can be attributed to the presence of considerable covalent character in the metal-sulfur bonds.

The *class b metals* are characterized by the presence of a number of *d* electrons beyond an inert-gas core. These *d* electrons can be used to π bond with ligand atoms, and the presence of such π bonding gives rise to many of the properties of the class *b* metals. The most stable complexes of these metals are formed with ligands that can accept electrons from the metal, that is, ligands with vacant *d* orbitals such as $P(CH_3)_3$, $S^=$, and I^-, or ligands with MO's into which electrons can be delocalized, such as CO and CN^- (Figure 2–21). It is therefore found that class *a* and *b* elements form stable compounds with different types of ligands. Class *a* elements prefer O- and N-containing ligands and F^-. Class *b* elements form more stable complexes with the heavier elements of the N, O, and F families.[1]

[1] Professor R. G. Pearson of Northwestern University [*J. Am. Chem. Soc.*, **85**, 3533 (1963)] has classified class *a* metals as *hard acids* and class *b* metals as *soft acids*. Ligand atoms such as N, O, and F are *hard bases* and those similar to P, S, and I are *soft bases*. The most stable complexes result from hard-hard acid-base and soft-soft acid-base combinations.

A completely adequate explanation of the stability of metal complexes is difficult, since the heat evolved in a complexation reaction (18) is small. A variety of relatively small effects such as

$$[M(H_2O)_x]^{n+} + yL \rightarrow [M(H_2O)_{x-y}L_y]^{n+} + yH_2O \qquad (18)$$

π bonding, crystal field stabilization, and increased covalent character in the metal-ligand bond can provide enough energy to alter the behavior that one might consider "normal."

Entropy changes also play an important role in determining complex stability. Reactions in which positive ions and negative ligands interact to form complexes of lower charge (19) proceed

$$[M(H_2O)_n]^{3+} + L^- \rightarrow [M(H_2O)_{n-1}L]^{2+} + H_2O \qquad (19)$$

with a large increase in entropy; this is a major factor in the stability of the resulting complex. The large entropy change arises mainly because each charged reactant has an ordered solvation sphere. The resulting products, having a lower charge, will produce considerably less ordering of the solvent. Fortunately, the factors that produce this increase in entropy are the same as those that produce stability in terms of electrostatics. Hence the electrostatic prediction of high stability due to the interaction of small highly charged ions may be accurate largely because of the entropy effect.

Entropy considerations are very important in two other ways. In the formation of a metal complex $[ML_6]^{n+}$ from $[M(H_2O)_6]^{n+}$, each replacement of an additional H_2O molecule by another ligand L becomes more difficult. For example, successive stepwise stability constants for the reaction (20) are $K_1 = 5 \times 10^2$, $K_2 = 1.3 \times 10^2$,

$$[Ni(H_2O)_6]^{2+} + 6NH_3 \rightleftharpoons [Ni(NH_3)_6]^{2+} + 6H_2O \qquad (20)$$

$K_3 = 4 \times 10^1$, $K_4 = 1.2 \times 10^1$, $K_5 = 4$, and $K_6 = 0.8$. This effect arises at least in part from the statistics of substitution processes (an entropy consideration). The replacement of one water by ammonia removes one possible site for the coordination of additional ammonia molecules. Moreover, the larger the number of ammonia molecules present in the complex, the greater the probability of their replacement by water. Both these factors reduce the probability of formation and hence the stability of the more highly substituted complexes. Other factors, such as steric repulsion between bulky ligands and electrostatic repulsion as anionic ligands replace water molecules on a positive metal ion, can also retard the coordination of additional ligands.

There are, however, a few examples in which the initial complexes are less stable than their more highly substituted relatives. Deviations from a regular decreasing trend of stepwise stability constants have been interpreted in some cases as indicating a change in the coordination number of the metal ion. The stability constants for $[CdBr_4]^{2-}$ are $K_1 = 2 \times 10^2$, $K_2 = 6$, $K_3 = 0.6$, $K_4 = 1.2$. Since the coordination numbers of cadmium in the hydrated ion and in $[CdBr_4]^{2-}$ are probably 6 and 4, respectively, the large value of K_4 may indicate that reaction (21), corresponding to this constant, in-

$$[Cd(H_2O)_3Br_3]^- + Br^- \rightleftharpoons [CdBr_4]^{2-} + 3H_2O \qquad (21)$$

volves a change in coordination number as well as the addition of Br^-. Therefore, because of the release of three molecules of water, this final step is accompanied by a large increase in entropy and hence a larger K_4.

The second very important entropy-induced effect is the great stability of metal chelates (see definition of chelate in Section 1–3). Both ammonia and ethylenediamine (en) coordinate with metals through amine nitrogens; in terms of the heat evolved in the complexation reaction two molecules of NH_3 have been shown to be about equivalent to one en. However, en complexes are considerably more stable than their NH_3 counterparts (for example, $[Ni(NH_3)_6]^{2+}$, $K_1K_2 = 6 \times 10^4$; $K_3K_4 = 5 \times 10^2$; $K_5K_6 = 3$. $[Ni(en)_3]^{2+}$, $K_1 = 2 \times 10^7$; $K_2 = 1.2 \times 10^6$; $K_3 = 1.6 \times 10^4$). It has been experimentally demonstrated that the unusual stability of the en compounds is due to a more favorable entropy associated with their formation.

Chelating ligands in general form more stable complexes than their monodentate analogs. This is known as the *chelate effect*, and it is explained in terms of the favorable entropy for the chelation process. In a qualitative way one can understand the favorable entropy. The replacement of a coordinated water molecule by either an NH_3 or en molecule should be about equally probable. The replacement of a second water molecule by the other amine group in the coordinated en, however, is much more probable than its replacement by a free NH_3 molecule from solution, since the en is already tied to the metal ion and the free end of the molecule is in the immediate vicinity of the H_2O it will replace. Thus the formation of $[Ni(en)(H_2O)_4]^{2+}$ is more probable than the formation of the less stable $[Ni(NH_3)_2(H_2O)_4]^{2+}$.

Another way to visualize the more favorable entropy change is to realize that a process in which the number of independent particles increases proceeds with an increase in entropy (the larger the number of particles, the greater is the possible disorder). In the coordination of one en molecule, two H_2O molecules are freed; this process should proceed with a favorable entropy.

Ter-, quadri-, and other polydentate ligands can replace three, four, or more coordinated water molecules, respectively, to form even more stable complexes. The stability constants for some complexes of Ni^{2+} with polydentate ligands are presented in Table 5-3. Ethylenediaminetetraacetate (EDTA), a sexadentate ligand (see XXVII in Section 3-4), forms stable complexes with a wide variety

TABLE 5-3

The Effect of Chelation on the Stability of Complexes[a]

Complex	β_1	β_2	β_3	β_4	β_5	β_6
$[Ni(NH_3)_6]^{2+}$	5×10^2	6×10^4	3×10^6	3×10^7	1.3×10^8	1.0×10^8
$[Ni(en)_3]^{2+}$	5×10^7	1.1×10^{14}	4×10^{18}			
$[Ni(dien)_2]^{2+b}$	6×10^{10}	8×10^{18}				
$[Ni(trien)(H_2O)_2]^{2+c}$	2×10^{14}					

[a] All β's were measured in 1 *M* KCl at 30°C. To see the effect of chelation, compare β's that are marked similarly. Each type of marking indicates complexes containing the same amount of coordinated H_2O.

[b] dien is $H_2\ddot{N}CH_2CH_2\ddot{N}HCH_2CH_2\ddot{N}H_2$.

[c] trien is $H_2\ddot{N}CH_2CH_2\ddot{N}HCH_2CH_2\ddot{N}HCH_2CH_2\ddot{N}H_2$.

of metal ions including the alkaline-earth metals (which form very unstable complexes with monodentate ligands). This compound is used commercially as a *sequestrant*, a reagent that will form complexes with metal ions and in this way control their concentration in solution. For example, EDTA will very efficiently complex calcium ion and is therefore an excellent water-softening agent. It is also used as an analytical reagent.

In basic solution EDTA reacts quantitatively with certain metals to produce the metal complex (22). It can therefore be used as a titrant for the volumetric determination of many metals. Several excellent indicators have been developed to detect the end point;

$$\mathrm{EDTA^{4-}} + [\mathrm{M(H_2O)_6}]^{2+} \rightarrow [\mathrm{MEDTA}]^{2-} + 6\mathrm{H_2O} \qquad (22)$$

such volumetric techniques are now common in quantitative chemical analysis.

Metal chelates contain rings of atoms, II. The stability of the

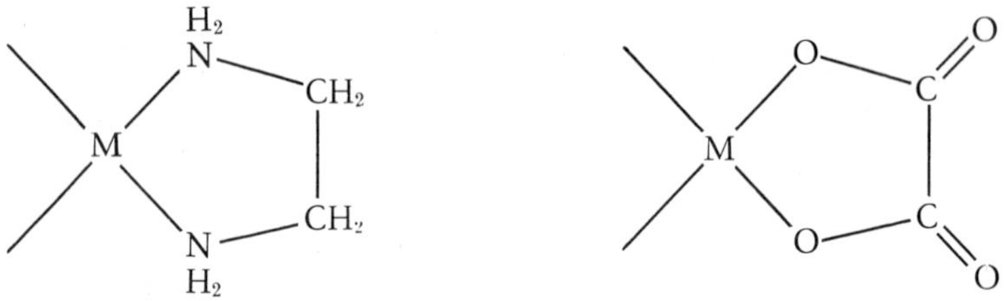

II

complex ion has been observed to be dependent on the number of atoms in the ring. In general it has been observed that for ligands that do not contain double bonds, those that form five-membered metal-chelate rings give the most stable products. Ligands that contain double bonds, such as acetylacetone, form very stable metal complexes containing six-membered rings. Chelate rings that contain either four atoms or more than six atoms have been observed, but they are relatively unstable and uncommon.

A very large amount of information on the stabilities of metal complexes is now available. This permits an assessment of the various factors that influence the stability of a metal complex. Some of these factors have been discussed in this chapter, but it may be helpful to summarize them briefly here. First, the stability of a complex obviously depends on the nature of the metal and of the ligand. With reference to the metal the following factors are important:

1. Size and charge. Because of the significant influence of electrostatic forces in these systems, the smaller the size and the larger the charge of a metal ion the more stable are the metal complexes. Thus stability is favored by a large charge-to-radius ratio of the metal ion.

2. Crystal field effects. The crystal field stabilization energy (CFSE) plays an important role in the stability of transition-metal complexes and appears to be responsible for the natural order of stability of the complexes of the first-row transition metals (Figure 5-2).

3. Class *a* and class *b* metals. The more electropositive metals, for example, Na, Ca, Al, the lanthanides, Ti, and Fe, belong to class *a*. The less electropositive metals, for example, Pt, Pd, Hg, Pb, and Rh, belong to class *b*. Class *a* metals form their most stable complexes with ligands in which the donor atom is N, O, or F; class *b* metals prefer ligands in which the donor atom is one of the heavier elements in the N, O, or F families. It is believed that the stability of the complexes of the class *b* metals results from an important covalent contribution to metal-ligand bonds and from the transfer of electron density from the metal to the ligand via π bonding.

With reference to the role of the ligands in determining the stability of metal complexes, the following factors are important:

1. Base strength. The greater the base strength of a ligand, the greater is the tendency of the ligand to form stable complexes with class *a* metals.

2. Chelate effect. The stability of a metal chelate is greater than that of an analogous nonchelated metal complex, for example, $[Ni(en)_3]^{2+} > [Ni(NH_3)_6]^{2+}$. The more extensive the chelation, the more stable the system; recall the very stable complexes of the sexadentate EDTA.

3. Chelate ring size. The most stable metal chelates contain saturated ligands that form five-membered chelate rings or unsaturated ligands that form six-membered rings.

4. Steric strain. Because of steric factors, large bulky ligands form less stable metal complexes than do analogous smaller ligands, for example, $H_2NCH_2CH_2NH_2$ forms more stable complexes than $(CH_3)_2NCH_2CH_2N(CH_3)_2$. The strain is sometimes due to the geometry of the ligand coupled with the stereochemistry of the metal complex. For example, $H_2NCH_2CH_2NHCH_2CH_2NHCH_2CH_2NH_2$ can coordinate its four nitrogens at the corners of a square, but $N(CH_2CH_2NH_2)_3$ cannot; thus the straight-chain tetramine forms more stable complexes with Cu^{2+} than does the branched-chain amine, which is unable to assume the preferred square planar geometry.

The consequences of charge, size, CF stabilization, etc. on the stability of metal complexes are vitally important to the chemistry of coordination compounds. For example, the oxidation potentials for metal ions change markedly as the type of ligand is changed (Table 5-4). As water is replaced by CN^-, EDTA, or NH_3 in complexes of Fe^{2+} and Co^{2+}, the tendency for oxidation to the M^{3+} state is markedly enhanced. These ligands form very much more stable complexes with M^{3+} ions than with M^{2+} ions; this provides the driving force for the oxidation. This is particularly true in Co^{2+} systems. The complex $[Co(H_2O)_6]^{3+}$ will oxidize H_2O to O_2; in contrast, aqueous solutions of Co^{2+} salts are readily oxidized by atmospheric

TABLE 5-4

The Oxidation Potentials for Some Cobalt and Iron Complexes

Reaction	*Oxidation potential, volts*
$[Fe(H_2O)_6]^{2+} \rightarrow [Fe(H_2O)_6]^{3+} + e$	−0.77
$[Fe(CN)_6]^{4-} \rightarrow [Fe(CN)_6]^{3-} + e$	−0.36
$[FeEDTA]^{2-} \rightarrow [FeEDTA]^{-} + e$	+0.12
$[Co(H_2O)_6]^{2+} \rightarrow [Co(H_2O)_6]^{3+} + e$	−1.84
$[Co(NH_3)_6]^{2+} \rightarrow [Co(NH_3)_6]^{3+} + e$	−0.10

O_2 to $[Co(III)L_6]$ complexes in the presence of ligands such as NH_3, CN^-, or NO_2^-. The large change in oxidation potential that results from the presence of these ligands is largely due to the fact that the ligands provide a greater CF than H_2O provides; this favors the conversion of the high-spin d^7 Co^{2+} complexes to the highly CF-stabilized low-spin Co^{3+} d^6 complexes.

5-3 DETERMINATION OF STABILITY CONSTANTS

The observations we have made in regard to the stability of metal complexes have come from the study of stability constant data. The experimental determination of stability constants is an important but often difficult task. Perhaps the greatest problem in equilibrium measurements is to determine which species are actually present in solution. Many studies in the past have been negated by more recent observations that indicate that species and equilibria had been neglected. Equilibrium constants have been measured by many different methods. Usually a solution of the metal ion and ligand is prepared, sufficient time is allowed for the system to come to equilibrium, and the concentrations of the species in solution are then measured.

It is apparent from the equilibrium expression (23) that if one

$$\begin{gathered} A + B \rightleftharpoons C \\ K = \frac{[C]}{[A][B]} \end{gathered} \tag{23}$$

measures the concentrations of A, B, and C present at equilibrium, the equilibrium constant can be calculated. In fact in the simple system (23), if one knows the amounts of A and B present prior to the formation of C and then can measure either A, B, or C at equilibrium, it is possible to calculate the concentrations of the other two species and to determine the equilibrium constant. In many systems involving metal complexes a variety of complex species exist in equilibrium. In these cases it may be necessary to make experimental measurements of the concentrations of more than one species. The ensuing calculations are sometimes tedious, although normally straightforward. For the more complicated systems computers are used to advantage.

The measurement of equilibrium concentrations of species is complicated by the fact that the measurement must not disturb the equilibrium. For example, in reaction (24) one could not measure

$$[Co(H_2O)_6]^{2+} + Cl^- \rightleftharpoons [Co(H_2O)_5Cl]^+ + H_2O \qquad (24)$$

the Cl^- concentration in solution by the precipitation of AgCl. The addition of Ag^+ would not only precipitate free chloride ion but also remove the Cl^- from the cobalt complex.

A second and difficult problem arises from the fact that equilibrium constants depend on activities rather than on concentrations. Since activities and concentrations are numerically equal in very dilute solutions, one can avoid the problem by keeping the concentrations of all species low. Unfortunately, this is seldom practical. Another approach is to determine stability constants in a series of solutions each containing different amounts of a "noncomplexing" salt such as $NaClO_4$. In this case the solution environment is altered from dilute "ideal" conditions largely by Na^+ ions and ClO_4^- ions, and deviations from unit activity coefficients are primarily due to these ions. By extrapolation to zero salt concentration, one effectively measures the stability constant in an environment where activity coefficients are unity.

Frequently, equilibrium data are collected in relatively concentrated solutions and no attempt is made to convert concentrations to activities. Equilibrium constants calculated by using concentrations rather than activities, called *concentration constants*, are quantitatively accurate only at the conditions under which they were measured. However, in most cases comparisons of concentration constants obtained at the same experimental conditions provide reliable information on the relative stabilities of analogous systems. It is important when using stability constants to realize that true stability constants can be used to calculate concentrations quantitatively only in solutions that are very dilute or other solutions in which activity coefficients are known. Concentration constants can be used to calculate concentrations of species but are quantitatively valid only under the experimental conditions used to determine the constants. Most frequently, therefore, one uses these constants in a semiquantitative fashion.

Concentrations of species in solution can be measured by a variety of methods that do not disturb the equilibrium under study. The most common are probably spectroscopy and electroanalysis. The former involves the absorption of light by the species to be studied; the latter involves the electrochemistry of the system being investigated. Spectroscopic techniques can be illustrated by studies on the Fe^{3+}—NCS^- equilibrium (25). Ferric ion and thiocyanate

$$Fe^{3+} + NCS^- \rightleftharpoons Fe(NCS)^{2+} \tag{25}$$

ion individually are almost colorless in solution. The $FeNCS^{2+}$ complex, however, exhibits a bright red-orange color; that is, the individual ions do not absorb visible light, whereas $FeNCS^{2+}$ does. The intensity of the red-orange color depends directly on the concentration of $FeNCS^{2+}$ and can be used to measure that concentration in solution. If one adds known amounts of Fe^{3+} and NCS^- to a solution and subsequently determines the amount of $FeNCS^{2+}$ formed by measuring the intensity of color in the solution, one can then calculate the $[Fe^{3+}]$ and $[NCS^-]$.

$$[Fe^{3+}] = [Fe^{3+}]_0 - [FeNCS^{2+}] \quad \text{and} \quad [NCS^-] = [NCS^-]_0 - [FeNCS^{2+}]$$

$[Fe^{3+}]_0$ and $[NCS^-]_0$ represent the concentration of Fe^{3+} and NCS^- prior to the formation of complex. The equilibrium constant can then be calculated by Equation (26). It should be pointed out that

$$K = \frac{[FeNCS^{2+}]}{[Fe^{3+}][NCS^-]} \tag{26}$$

this measurement is actually more complicated than the description makes it appear, since species such as $[Fe(NCS)_2]^+$ also exist. It is possible to do the experiment with an excess of Fe^{3+} ion so that the concentrations of higher complexes are negligible and essentially the only species present are Fe^{3+} and $FeNCS^{2+}$.

The simplest electroanalytical technique for the determination of stability constants is one that makes use of the glass electrode. This device is the fundamental component of the common laboratory pH meter used to determine the activity of H^+ in solution; therefore, equilibria studied with the device must involve changes in $[H^+]$. Professor J. Bjerrum, of the University of Copenhagen, when a student determined the stability constants of a variety of NH_3 complexes by using this technique. The NH_3 concentration in a solution is related to the $[H^+]$ by the equilibrium constant for the acid dissociation of NH_4^+ (27). In acidic solution the $[NH_4^+]$ is large with

$$NH_4^+ \rightleftharpoons NH_3 + H^+$$

$$K = \frac{[NH_3][H^+]}{[NH_4^+]} \qquad [NH_3] = K[NH_4^+]\frac{1}{[H^+]} \tag{27}$$

respect to $[NH_3]$; thus $[NH_4^+]$ is virtually unchanged as the $[H^+]$ is increased. In these acid solutions

$$[NH_3] = K[NH_4^+]\frac{1}{[H^+]} = K'\frac{1}{[H^+]}$$

Therefore, under appropriate conditions the glass electrode that detects $[H^+]$ can also measure $[NH_3]$ directly. Stability constants for the Ag^+-NH_3 system were determined by using this technique (28), (29). Solutions were prepared at 30.0°C; all contained various

$$Ag^+ + NH_3 \rightleftharpoons [Ag(NH_3)]^+ \qquad K_1 \tag{28}$$

$$[Ag(NH_3)]^+ + NH_3 \rightleftharpoons [Ag(NH_3)_2]^+ \qquad K_2 \tag{29}$$

small concentrations of NH_3 and Ag^+. The pH of each of these solutions was measured and from this the concentration of uncomplexed $NH_3([NH_3])$ was calculated by using Equation (27). Some of the data are presented in Table 5–5. The average number of NH_3 molecules bound per $Ag^+(\bar{n})$ can be determined from the experimental data by means of Equation (30). A plot of $\bar{n}$ vs. NH_3 is shown in Fig-

$$\bar{n} = \frac{[NH_3]_0 - [NH_3]}{[Ag^+]} \tag{30}$$

ure 5–3. This plot shows that at concentrations of free ammonia greater than 10^{-2} M, $[Ag(NH_3)_2]^+$ is the predominant species,

TABLE 5-5

pH Data Used to Determine the Stability Constants for Species in the Aqueous Ag^+-NH_3 System[a]

$[Ag^+]_0$	$[NH_3]_0$	pH	$[H^+]$	$[NH_3]$	$\bar{n}$
0.0200	0.00502	4.970	10.7×10^{-6}	0.88×10^{-4}	0.246
0.0200	0.01504	5.372	4.25×10^{-6}	2.21×10^{-4}	0.740
0.0200	0.03012	5.793	1.61×10^{-6}	5.83×10^{-4}	1.477
0.0200	0.05022	6.342	0.455×10^{-6}	20.6×10^{-4}	1.895

[a] The data were taken at 30°C in the presence of 2 *M* NH_4NO_3.

whereas at free NH_3 concentrations below 10^{-4} *M*, Ag^+ is the abundant species. The complex $[Ag(NH_3)]^+$ is present within only a small $[NH_3]$ range. From these experimental data the stability constants $K_1 = 2.5 \times 10^3$ and $K_2 = 8.3 \times 10^3$ were calculated. The calculations are rather tedious and beyond the scope of this text. Note that this is another example in which the second stepwise con-

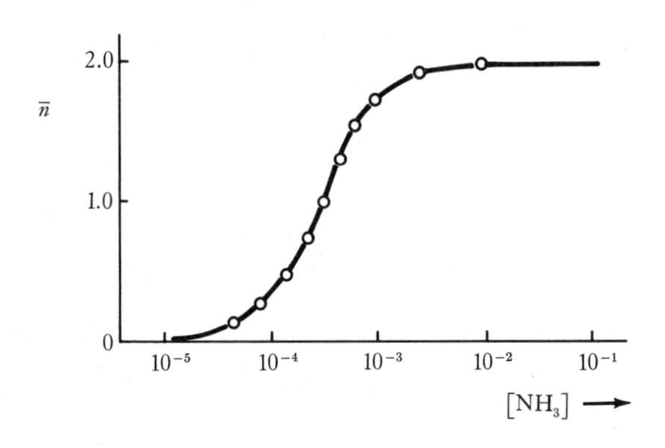

Figure 5-3 The average number of NH_3 molecules bound per Ag^+ in solutions containing a variety of concentrations of free ammonia. The data apply to solutions at 30°C containing 2.0 *M* NH_4NO_3.

stant is greater than the first and is an exception to the general rule that $K_1 > K_2 > K_3 > \cdots$ (Section 5–2).

Two methods for determining stability constants have been briefly described. Many other experimental methods are available. For example, methods that make use of radioactive isotopes or employ liquid-liquid extraction or ion exchange are fairly common. Virtually any technique that can be used to determine concentration can be and has been used to determine stability constants.

The stability of metal complexes in solution is one important aspect of the solution chemistry of metals. The structure of the solvent, the formulas and nature of the hydration spheres of the solute species present, and the reactions between species and their neighbors and the equilibria resulting therefrom have interested scientists for many years. Large volumes of information have been gathered, and elaborate theories have been developed to explain the experimental data. In spite of all this effort, solution chemistry remains a challenging research field. Basic detailed information, such as the number of water molecules that surround most aqueous ions and reliable stability constants for many species, especially the unstable and the very stable, are not yet available.

PROBLEMS

1. (*a*) Write all the stepwise (K's) and overall (β's) stability constants for the following reactions. (*b*) Estimate which of all the stepwise constants will be the greatest. (*c*) Predict which of all the stepwise constants will be the smallest.

$$Ni^{2+} + 4CN^- \rightleftharpoons [Ni(CN)_4]^{2-} \text{(diamagnetic)}$$
$$Ag^+ + 2NH_3 \rightleftharpoons [Ag(NH_3)_2]^+$$
$$Cr^{3+} + 3en \rightleftharpoons [Cr(en)_3]^{3+}$$
$$Fe^{3+} + 4Cl^- \rightleftharpoons [FeCl_4]^-$$

2. The following compounds were dissolved in 100cc of water: 1.00×10^{-3} moles of dien ($H_2NCH_2CH_2NHCH_2CH_2NH_2$) and 5.00×10^{-3} moles of $Ni(ClO_4)_2$. (*a*) Calculate the concentration of $[Ni(dien)]^{2+}$ in solution (you can assume that the concentration of $[Ni(dien)_2]^{2+}$ is much less than either Ni^{2+} or $[Ni(dien)]^{2+}$). (*b*) Calculate the concentration of Ni^{2+} in solution. (*c*) Calculate the concentration of $[Ni(dien)_2]^{2+}$ in solution.

$$K_1 = 5.0 \times 10^{10} \qquad K_2 = 1.6 \times 10^8$$

3. One learns from qualitative analysis experiments that AgCl dissolves in an excess of aqueous NH_3, whereas AgI does not. This result is due to the stability of $[Ag(NH_3)_2]^+$, which is sufficiently large to make the AgCl dissolve but not large enough to dissolve the less soluble AgI. A solution containing 0.15 *M* Cl^- and 0.15 *M* I^- is made 5 *M* in NH_3; then solid $AgNO_3$ is added in an amount equivalent to the $Cl^- + I^-$ concentrations. Calculate if AgCl and/or AgI will precipitate.

$$K_{sp}(AgCl) = 1.7 \times 10^{-10}$$
$$K_{sp}(AgI) = 8.5 \times 10^{-17}$$
$$\beta_2([Ag(NH_3)_2]^+) = 1.5 \times 10^7$$

4. (*a*) Calculate whether PbS would precipitate from a solution containing 0.5 *M* $EDTA^{4-}$, 0.001 *M* S^{2-}, and 0.01 *M* Pb^{2+}. (*b*) Make the same calculations for Ni^{2+}, Co^{2+}, Zn^{2+}, and Cd^{2+}.

Cation	$K\,[M(EDTA)]^{2-}$	$K_{sp}[MS]$
Pb^{2+}	2×10^{18}	4×10^{-26}
Ni^{2+}	3.6×10^{18}	1×10^{-22}
Co^{2+}	1.6×10^{16}	5×10^{-22}
Zn^{2+}	3.9×10^{16}	6×10^{-27}
Cd^{2+}	2.6×10^{16}	1×10^{-20}

REFERENCES

J. Bjerrum, *Metal Ammine Formation in Aqueous Solution*, Haase, Copenhagen, 1941.

A. E. Martell and M. Calvin, *Chemistry of the Metal Chelate Compounds*, Prentice-Hall, Englewood Cliffs, N.J., 1952.

S. Chaberek and A. E. Martell, *Sequestering Agents*, Wiley-Interscience, New York, 1959.

F. J. C. Rossotti and H. Rossotti, *The Determination of Stability Constants*, McGraw-Hill, New York, 1961.

Data on the stability constants of metal complexes are continually reported in the scientific literature. A few years ago the available data were collected in the form of tables and published in two volumes. Volume three of this series is about to appear.

J. Bjerrum, G. Schwarzenbach, and L. G. Sillen (eds.), *Stability Constants of Metal-Ion Complexes:* Part I, *Organic Ligands;* Part II, *Inorganic Ligands*, Chemical Society of London, 1957, 1958.

VI

Kinetics and Mechanisms of Reactions of Coordination Compounds

Perhaps the most important use for metal complexes is in the catalysis of reactions. Studies of metal enzymes (physiological catalysts) show that the site of reaction in the biological system is frequently a complexed metal ion. Several industrial processes depend directly on catalysis by metal complexes. Mention was made in Chapter I of the production of polyethylene by a process using a complex of aluminum and titanium as a catalyst. The reaction of an olefin with carbon monoxide and hydrogen takes place in the presence of a cobalt complex (1). This very important *oxo* reaction has

$$CH_3CH{=}CH_2 + CO + H_2 \xrightarrow{Co} CH_3CH_2CH_2CHO \qquad (1)$$

been studied in detail; the catalyst is known to be $HCo(CO)_4$, which is generated in the reaction. The air oxidation of ethylene to produce acetaldehyde takes place readily in the presence of a $PdCl_2$–$CuCl_2$ catalyst system (2). This very recently developed in-

$$CH_2{=}CH_2 + \tfrac{1}{2}O_2 \xrightarrow[CuCl_2]{PdCl_2} CH_3CHO \qquad (2)$$

dustrial process, the *Wacker* process, depends on the intermediate formation of the complex $[Pd(C_2H_4)(OH)Cl_2]^-$. These and many other recent applications of metal complexes are exciting the imagination of research chemists and increasing the production potential and versatility of our chemical industry. In order to exploit the use of metal complexes, it is necessary that we know more about some of the details of the reaction processes. This chapter illustrates the approach to such problems and provides examples of information obtained and reaction theories proposed.

In previous chapters we have looked at a variety of reactions of coordination compounds. Some of these reactions produced coordination compounds from simpler species; others transformed one coordination compound into another. In Chapter V we found that the equilibrium constants for these reactions depend on the heat evolved and the amount of disorder produced (entropy). A favorable heat or entropy change is a necessary condition for the occurrence of a reaction. However, the reaction rate must also be sufficiently fast in order for a reaction to proceed. Reactions occur at a variety of speeds; some are immeasurably slow, and others are so rapid that only very recently has it been possible to measure their rates.

Some reactions, such as the very exothermic combination of H_2 and O_2 to give H_2O, do not occur until the mixture is ignited. Other, less exothermic reactions, such as the endothermic solution of salt in water, proceed rapidly at room temperature. This indicates that the rate of a reaction does not necessarily depend on the magnitude of the heat of the reaction. Reactions with very favorable equilibrium constants need not occur at a rapid rate. *The rate of a chemical reaction is dependent on the nature of the path by which the reactants are transformed into products (the mechanism of the reaction).* Knowing the mechanism of a reaction often makes it possible to understand the rate behavior of the reaction. What is more important in practice, it is possible to learn a great deal about the reaction mechanism from the rate behavior of the reaction.

6–1 RATE OF A REACTION

The rate of a reaction such as (3) can be expressed as the decrease in the number of moles of reactants, $[Co(NH_3)_5Cl]^{2+}$ and H_2O, per

$$[Co(NH_3)_5Cl]^{2+} + H_2O \rightarrow [Co(NH_3)_5H_2O]^{3+} + Cl^- \qquad (3)$$

second (or some other unit of time), or the increase in the number of moles of products, $[Co(NH_3)_5H_2O]^{3+}$ or Cl^- per second. Since the disappearance of 1 mole of $[Co(NH_3)_5Cl]^{2+}$ produces 1 mole of $[Co(NH_3)_5H_2O]^{3+}$ and 1 mole of Cl^-, these three rates would be numerically equal. In general, the *rate of* any *reaction* can be defined as the *change in concentration of any of the reactants or products per unit of time*.

A convenient way of expressing a rate quantitatively is in terms of half-life. The *half-life* of a reaction is the amount of time necessary for one-half of a reactant to be consumed or the time for one-half of a product to be formed. The half-life of reaction (3) at 25°C has been found to be 113 hr. This means that if one dissolved a salt containing $[Co(NH_3)_5Cl]^{2+}$ in water at 25.0°C, after 113 hr only one-half of the $[Co(NH_3)_5Cl]^{2+}$ would remain, half having been converted to $[Co(NH_3)_5H_2O]^{3+}$ and Cl^- (Figure 6–1). After the next 113 hr one-half of the remaining $[Co(NH_3)_5Cl]^{2+}$ will have reacted, leaving only one-fourth of the original amount, and so on. Although H_2O is a

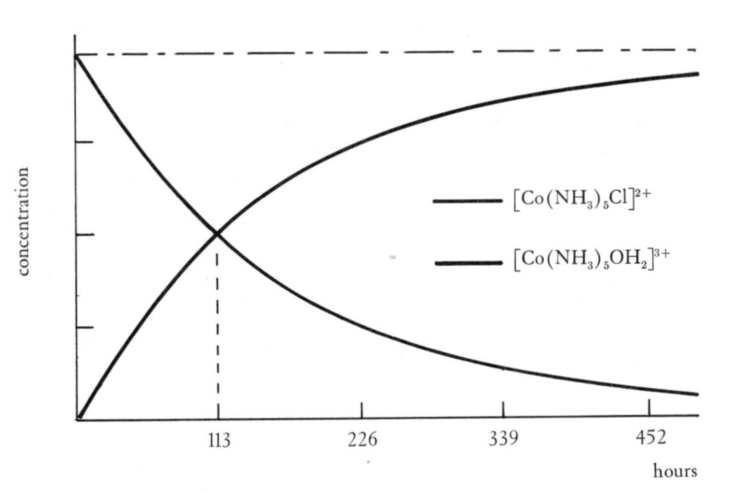

Figure 6–1 A plot of the concentrations of reactants and products of reaction (3) as a function of time at 25°C.

reactant in this reaction, its concentration would not have been halved in the first 113 hr, since, as the solvent, it is present in very large excess. The concentrations of $[Co(NH_3)_5H_2O]^{3+}$ and Cl^- will be one-half of the values they will achieve when the reaction is complete.

6–2 THE RATE LAW

Having defined the rate of a reaction, let us now see how one can learn something about rate phenomena from the consideration of reaction mechanisms. The simplest type of reaction one can visualize is the isomerization (4) or dissociation (5) of a molecule. Reactions

$$A \rightarrow A' \tag{4}$$

$$A \rightarrow B + C \tag{5}$$

of this stoichiometry can occur via complicated mechanisms in which a variety of intermediate products are formed; however, let us first choose the simplest mechanism, one in which A at some instant goes directly to A′ (or B + C). One would expect under these circumstances that the rate of the reaction should depend on the concentration of A and no other species. The more molecules of A present, the greater will be the probability for one molecule to react. Thus the rate of reaction is directly proportional to the concentration of A (6). This can be written by using a constant k that is

$$\text{rate} \propto [A] \tag{6}$$

called the *rate constant* and is a number characteristic of the reaction rate for a given temperature (7). For reactions that proceed rapidly,

$$\text{rate} = k[A] \tag{7}$$

k is a large number; for slow reactions, k is small. There are a variety of reactions for which a simple rate expression of this type applies. An example is the conversion of *cis*-$[Co(en)_2Cl_2]^+$ to *trans*-$[Co(en)_2Cl_2]^+$ in methanol solution (8). The rate of conversion of

$$cis\text{-}[Co(en)_2Cl_2]^+ \xrightarrow{k} trans\text{-}[Co(en)_2Cl_2]^+ \qquad (8)$$

the cis to the trans isomer is equal to the product of the rate constant for the reaction and the concentration of the cis isomer (9).

$$\text{rate} = k[cis\text{-}Co(en)_2Cl_2^+] \qquad (9)$$

The reaction may also proceed by a more complicated mechanism (10). In this mechanism A is converted to A′ by a process involving

$$A + D \xrightarrow{\text{slow}} E \quad (a) \qquad E \xrightarrow{\text{fast}} A' + D \quad (b) \qquad (10)$$

the initial formation of an intermediate E (10*a*) and its subsequent conversion to A′ (10*b*). The formation of E must involve a collision of A and D. The rate of this process should be proportional to the concentration of A and D (11), since the probability of a collision

$$\text{rate of formation of E} \propto [A][D] \qquad (11)$$

depends directly on these concentrations. In a multistep process, the over-all reaction rate is determined by the slow step that is called the *rate-determining step*. If the dissociation of E is much faster than its formation, A′ will be formed effectively as rapidly as E. Then the rate of formation of A′ is equal to the rate of formation of E. Rewriting the expression for E, we find the rate of formation of A′ is given by expression (12). D is not consumed in the reaction, and yet

$$\text{rate} = k[A][D] \qquad (12)$$

the rate is dependent on its concentration; it is called a *catalyst*. The interconversion (13) of the optical isomers of $[Co(en)_3]^{3+}$ is catalyzed by $[Co(en)_3]^{2+}$, and the rate expression for the reaction has the form (14). The rate-determining step in reaction (13) is known to involve the transfer of an electron from $[Co(en)_3]^{2+}$ to $[Co(en)_3]^{3+}$ (Section 6–8).

A third but most unlikely mechanism for the reaction is that shown by (15), (16). This mechanism involves the slow formation

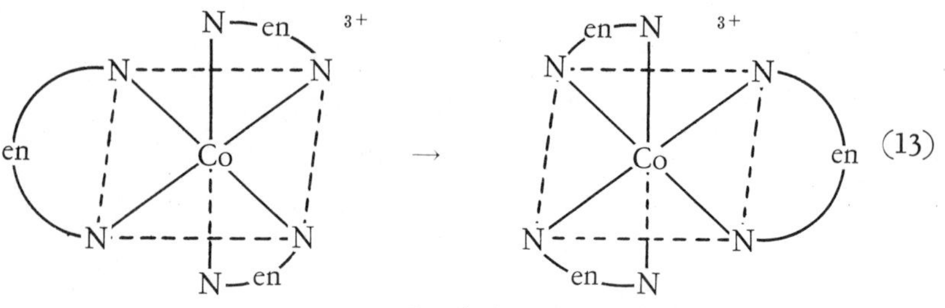

$$\text{rate} = k[\text{Co(en)}_3{}^{3+}][\text{Co(en)}_3{}^{2+}] \tag{14}$$

$$\text{A} + \text{D} + \text{D} \xrightarrow{\text{slow}} \text{E} \tag{15}$$

$$\text{E} \xrightarrow{\text{fast}} \text{A}' + 2\text{D} \tag{16}$$

of an intermediate E by the collision of A and two molecules of D (15). The rate of the formation of E, and of A′, if E decomposes as rapidly as it is formed, is given by the expression (17). A collision of

$$\text{rate} = k[\text{A}][\text{D}][\text{D}] = k[\text{A}][\text{D}]^2 \tag{17}$$

three bodies is very improbable; hence, reactions that proceed by this type of process are very slow and very rare.

The rate expressions that were written for the three different paths from A to A′ are called *rate laws*. They describe the effect of concentration on the rate of a reaction. A *first-order* rate law has the form (7); the reaction rate (or simply the reaction) is said to have a *first-order dependence* on A or to be first order in [A]. Reactions that obey the *second-order rate law* (12) are said to have a first-order dependence on both[A] and [D]. The *third-order rate law* (17) indicates that reactions that obey this law will have a first-order rate dependence on [A] and a *second-order* dependence on [D].

The order of a reaction depends on the number of species and the number of times each species appears in the rate law. Often, the order of a reaction is equal to the number of particles that collide in the rate-determining step of the reaction. Later, however, we shall have examples in which the order of a reaction is less than the number of particles involved in the rate-determining step. From this discussion it should be apparent that the rate law for a reaction cannot be determined from stoichiometry. The rate law for the reaction $A \rightarrow A'$ may contain a variety of species not included in the chemical reaction; also, it need not include A or A′. If one can determine experimentally the rate law for a reaction, one may learn which species are involved in the rate-determining step and hence obtain vital information about the reaction mechanism.

6–3 EFFECTIVE COLLISIONS

If it were possible to predict rate constants, then it would be possible to determine which reactions would proceed at a rapid rate and which would be impractically slow. A theoretical approach to rate constants is available in collision theory. The rate of a reaction is given by a rate law that consists of a rate constant and the concentrations of the species involved in the slow step of the reaction. The concentration dependence of the rate law is related to the probability that collision between reacting species will occur. If each collision led to a reaction, the rate constant would play a trivial role. In fact, in most reactions many collisions are ineffective; the rate constant is a measure of the collision effectiveness, and its magnitude stems primarily from the geometry and violence required in the collision.

The geometry of a collision must be suitable. For reactions of particles other than spherical molecules or ions, the particles must collide with a definite orientation for reaction to occur. For example, a cyanide ion must approach a metal ion with its carbon end in order for a metal-carbon bond to form (18). Reaction geometry is an

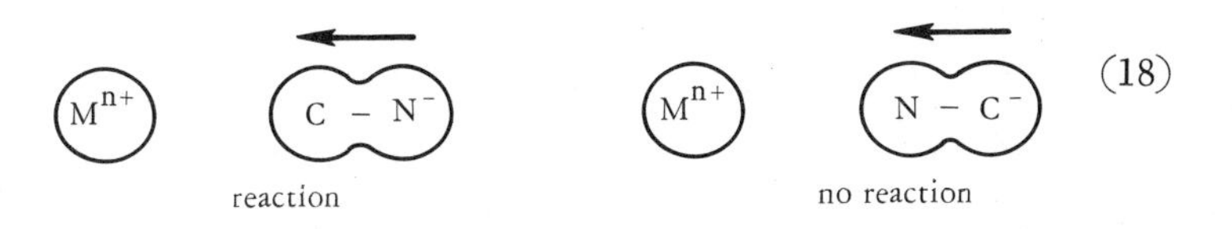

important factor in gas-phase reactions, but it is less important in solution. Molecules in solution are effectively held in a cage by neighboring solvent molecules, and hence they normally collide with their neighbors many times before moving on to another site. Thus whenever a CN^- comes near a metal ion, it will collide with it many times before it escapes from its cage; almost certainly some of these collisions will have the proper orientation for reaction.

The most important rate-determining factor in most reactions is the *collision energy*. In the reaction of ammonia with aqueous Ag^+, the NH_3 molecule must take the place of a coordinated H_2O molecule. The collision must provide the necessary energy for this process; otherwise, the reaction does not occur. When molecules have used their collisional energy to come to a configuration such that the reaction will proceed without further addition of energy, they are said to be in the *activated complex*. The amount of energy necessary to form the activated complex is called the *activation energy* (Figure 6–2). In reactions that have small activation energies most collisions will be sufficiently energetic to lead to reaction. A very high activation energy will make all but the most violent collisions ineffective. The magnitude of the rate constant for a reaction in general inversely parallels the magnitude of the activation energy. The mechanism of the reaction determines the configuration and energy of the activated complex and hence the activation energy and rate of the reaction.

Reactions having large activation energies can be made to proceed at more convenient rates by increasing the reaction temperature or by using a catalyst. An increase in temperature increases the velocity of the reactant particles and hence the violence of their

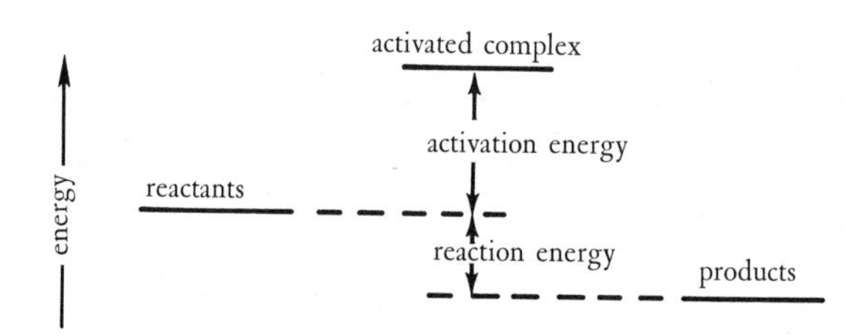

Figure 6–2 The relative energies of reactants, activated complex, and products of a reaction.

collisions. On the other hand, catalysts so change the reaction mechanism that the new activated complex in which the catalyst is present can be formed by less energetic collisions.

A reaction can also occur by a mechanism that does not involve a collision. In such a reaction collision geometry does not, of course, affect the rate constant. However, the simple reactions of this type usually have an activation energy. The reacting molecule must store up energy from collisions with neighbors (solvent molecules) or by absorbing radiation in order to attain the configuration of the activated complex; then reaction can occur. The rate constant for this type of process will be a measure of how often a molecule will collect enough energy to react.

It is possible to visualize a variety of mechanisms for all reactions; the one that is observed is the one that produces the fastest reaction under the condition of the experiment. Other processes that are slower will make small or negligible contributions to the total reaction.

6–4 INERT AND LABILE COMPLEXES

Complexes in which ligands are rapidly replaced by others are called *labile complexes;* those in which ligand substitution is slow are called *inert*. To make this distinction somewhat quantitative, Henry Taube, now professor of chemistry at Stanford University, suggested that those complexes in which the substitution of ligands takes place in less than one minute be called labile. The reaction conditions are specified as a temperature of 25°C and 0.1 *M* reactants. Although it is often found that a stable complex is inert and that an unstable complex is labile, no such correlation is required. Cyanide ion forms very stable complexes with metal ions such as Ni^{2+} and Hg^{2+}. The stability indicates that equilibrium (19) lies far to the

$$[Ni(H_2O)_6]^{2+} + 4CN^- \rightleftharpoons [Ni(CN)_4]^{2-} + 6H_2O \qquad (19)$$

right and that Ni^{2+} prefers CN^- to H_2O as a ligand. However, when ^{14}C-labeled cyanide ion is added to the solution, it is almost instantaneously incorporated into the complex[1](20). Thus the stability of this complex does not ensure inertness.

$$[Ni(CN)_4]^{2-} + 4\,^{14}CN^- \rightleftharpoons [Ni(^{14}CN)_4]^{2-} + 4CN^- \qquad (20)$$

Cobalt(III) ammine complexes such as $[Co(NH_3)_6]^{3+}$ are unstable in acid solution. At equilibrium almost complete conversion to $[Co(H_2O)_6]^{2+}$, NH_4^+, and O_2 is observed (21). However,

$$4[Co(NH_3)_6]^{3+} + 20H^+ + 26H_2O \rightarrow 4[Co(H_2O)_6]^{2+} + 24NH_4^+ + O_2 \qquad (21)$$

$[Co(NH_3)_6]^{3+}$ can be kept in acid solution for days at room temperature without noticeable decomposition. The rate of decomposition is very low; hence the compound is unstable in acid solution but inert.

In Chapter V the stability of coordination compounds was discussed; in this chapter reaction rate or lability is considered. It is important to remember that these terms relate to different phenomena. The *stability* of a complex depends on the *difference in energy between reactants and products* (the reaction energy in Figure 6–2). A stable compound will be considerably lower in energy than possible products. The *lability* of a compound depends on the *difference in energy between the compound and the activated complex;* if this activation energy is large, the reaction will be slow.

For six-coordinated complexes it is possible to predict with some degree of reliability which are labile and which are inert. Taube first called attention to this by pointing out that the electronic structure of a complex plays a significant role in the rate of its reactions. A classification of six-coordinated complexes in terms of the number and type of *d* electrons present in the central atom follows.

Labile Complexes

1. All complexes in which the central metal atom contains *d* electrons in e_g orbitals (the $d_{x^2-y^2}$ and d_{z^2} orbitals that point toward

[1] The labeled $^{14}CN^-$ is virtually identical chemically with unlabeled CN^-; therefore the reaction proceeds until the ratio $^{14}CN^-/CN^-$ in the complex is equal to that in solution.

the six ligands; see Section 2–5), for example, $[Ga(C_2O_4)_3]^{3-}$, d^{10} $(t_{2g}{}^6e_g{}^4)$; $[Co(NH_3)_6]^{2+}$, d^7 $(t_{2g}{}^5e_g{}^2)$; $[Cu(H_2O)_6]^{2+}$, d^9 $(t_{2g}{}^6e_g{}^3)$; $[Ni(H_2O)_6]^{2+}$, d^8 $(t_{2g}{}^6e_g{}^2)$; $[Fe(H_2O)_6]^{3+}$, d^5 $(t_{2g}{}^3e_g{}^2)$.

2. All complexes that contain less than three *d* electrons, for example, $[Ti(H_2O)_6]^{3+}$, d^1; $[V(phen)_3]^{3+}$, d^2; $[CaEDTA]^{2-}$, d^0.

Inert Complexes

Octahedral d^3 complexes, plus low-spin d^4, d^5, and d^6 systems, for example, $[Cr(H_2O)_6]^{3+}$, d^3 $(t_{2g}{}^3)$; $[Fe(CN)_6]^{3-}$, d^5 $(t_{2g}{}^5)$; $[Co(NO_2)_6]^{3-}$, d^6 $(t_{2g}{}^6)$; $[PtCl_6]^{2-}$, d^6 $(t_{2g}{}^6)$.

By using this classification, one can predict whether an octahedral complex is inert or labile if one knows its magnetic properties (whether it is high- or low-spin) and the number of *d* electrons present in the central atom.

The use of crystal field theory makes it possible to present a more detailed classification than simply "inert" and "labile." The approach depends on a comparison of the CFSE of a coordination compound and of its activated complex (recall that "activated complex" refers to a configuration of reactant molecules such that the reaction can proceed without further addition of energy).

If the CFSE of the compound is much larger than that of the activated complex, the compound will react slowly; if the difference is small, the reaction will be rapid. The difference between the CFSE of a compound and that of an activated complex derived from the compound affects the rate of a reaction because the change in CFSE is added to the activation energy for the process. If the activated complex is less CF stabilized than the original compound, the loss in stability in going to the activated complex increases the activation energy for the reaction and hence decreases the rate.

Calculations have been made of the CFSE's of octahedral complexes and for square pyramidal activated complexes (Table 6–1). From these data one can calculate the loss in CFSE on formation of the activated complex. There is considerable evidence that many octahedral complexes react by a process involving a five-coordinated intermediate (Section 6–7); however, these calculations must be taken as a crude approximation, since the model on which they are based may not be strictly valid in any reaction and is certainly basically incorrect in some.

TABLE 6–1

Crystal field stabilization energies for octahedral and square-pyramidal high-spin complexes

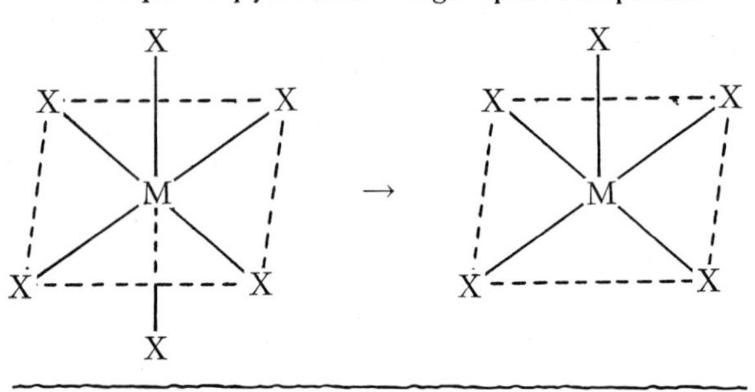

System	*CFSE, Δ* Octahedral	Square pyramid	*Change in CFSE,* Δ_o
d^0	0	0	0
d^1,d^6	0.40	0.45	−0.05
d^2,d^7	0.80	0.91	−0.11
d^3,d^8	1.20	1.00	+0.20
d^4,d^9	0.60	0.91	−0.31
d^5,d^{10}	0		0

The data in Table 6–1 show that there is an appreciable loss in CFSE in going to a square pyramidal activated complex from an octahedral d^3 or d^8 complex. These complexes are thus expected to react slowly, and indeed they do. All other high-spin complexes are expected to react rapidly, and they have been observed to do so. Similar calculations have been made for low-spin complexes. By using both the high- and low-spin calculations it appears that the rates of reactions of similar inert complexes should decrease in the order $d^5 > d^4 > d^8 \sim d^3 > d^6$ (the d^5, d^4, and d^6 systems are low-spin configurations). There has been some experimental confirmation for this sequence.

It is also possible to predict in more detail the rate behavior of complexes by considering the charge and size of their central atoms. The rules that were used to explain the stability of metal complexes (Section 5–2) often apply to kinetic behavior as well. Small highly charged ions form the most stable complexes; likewise, these ions form complexes that react slowly. Thus there is a decrease in lability with increasing charge of the central atom for the isoelectronic series $[AlF_6]^{3-} > [SiF_6]^{2-} > [PF_6]^{-} > SF_6$. Similarly, the rate of water exchange (22) decreases with increasing cationic charge in the

$$[M(H_2O)_6]^{n+} + 6H_2O^* \rightleftharpoons [M(H_2O^*)_6]^{n+} + 6H_2O \qquad (22)$$

order $[Na(H_2O)_n]^{+} > [Mg(H_2O)_n]^{2+} > [Al(H_2O)_6]^{3+}$.

Complexes having central atoms with small ionic radii react more slowly than those having larger central ions, for example, $[Mg(H_2O)_6]^{2+} < [Ca(H_2O)_6]^{2+} < [Sr(H_2O)_6]^{2+}$. For a series of octahedral metal complexes containing the same ligands, the complexes having central metal ions with the largest charge-to-radius ratios will react the slowest. The validity of this generalization is supported by the water exchange rate data summarized in Figure 6–3. It is of interest to note that of the first-row transition elements in Figure 6–3, the slowest to react is $[Ni(H_2O)_6]^{2+}$, the d^8 system, as is predicted by CFT. (The hydrated M^{2+} ions of the first-row transition elements are all high-spin complexes.) The rapid rate for $[Cu(H_2O)_6]^{2+}$ has been attributed to the exchange of water molecules above and below the square plane of the tetragonally distorted octahedral complex. The four water molecules in the square plane appear to react at a considerably slower rate.

In general, four-coordinated complexes (both tetrahedral and square planar molecules) react more rapidly than analogous six-coordinated systems. As was stated earlier, the very stable $[Ni(CN)_4]^{2-}$ undergoes rapid exchange with $^{14}CN^-$ (20). The rate of exchange is slow for six-coordinated complexes having about the same stability, for example, $[Mn(CN)_6]^{4-}$ and $[Co(CN)_6]^{3-}$. The greater rapidity of reactions of four-coordinated complexes may be due to the fact that there is enough room around the central ion for a fifth group to enter the coordination sphere. The presence of an additional group would aid in the release of one of the original ligands.

$$[M(H_2O)_x]^{n+} + x{*H_2O} \rightleftharpoons [M(*H_2O)_x]^{n+} + xH_2O$$

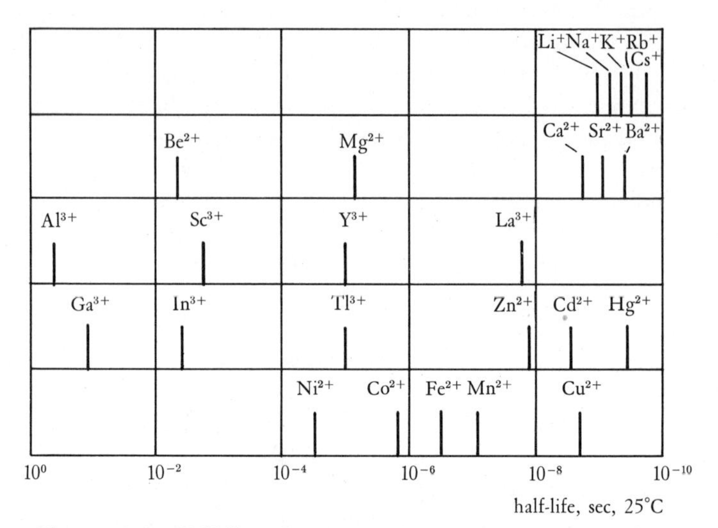

Figure 6–3 Half-lives for the exchange of water with hydrated metal ions. [Taken from M. Eigen, *Pure Appl. Chem.*, 6, 105 (1963).]

For square planar complexes it is not possible to apply successfully the charge-to-radius ratio generalization that works well for six-coordinated complexes. Thus for the nickel triad the size of the M^{2+} ions increases with increasing atomic number, but the rate of reaction decreases in the order $Ni^{2+} > Pd^{2+} >> Pt^{2+}$. The rate of $*Cl^-$ exchange by $[AuCl_4]^-$ is approximately 10^4 times faster than that by $[PtCl_4]^{2-}$, although the reverse behavior is expected in terms of the charges on the metal ions.

As was noted earlier, the *rate* of a reaction *depends upon* the *mechanism;* this involves the configuration and energy of the activated complex and hence the activation energy. For *octahedral systems* the activation energy is strongly influenced by the *breaking of metal-ligand bonds;* therefore, a large *positive charge* on the central atom *retards* the *loss of a ligand*. In *four-coordinated systems* the *formation of new* metal-ligand *bonds* is of increased importance, and it is *favored by* a large *positive charge on* the *metal* ion.

Therefore, the rules that predict rate behavior for six-coordinated systems will often not apply to complexes having smaller coordination numbers. Since rate behavior is dependent on mechanism and since reactions of metal complexes are known to proceed by a variety of paths, it is impossible to make generalizations that apply to all complexes regardless of the type of mechanism by which they react. In spite of this, the rules that are outlined in this section are surprisingly consistent with the data on the rate behavior of octahedral complexes.

6–5 MECHANISMS OF SUBSTITUTION REACTIONS

Now let us consider the application of kinetic and other techniques to the determination of reaction mechanisms. Reactions of coordination compounds can be divided into two broad categories: *substitution reactions* and *redox reactions*. In each of these a variety of mechanistic paths may be operable.

There are two basic mechanisms for substitution reactions: *dissociative* and *displacement* processes. We can illustrate these two mechanisms for the general octahedral substitution reaction (23).

$$[MX_5Y] + Z \rightarrow [MX_5Z] + Y \qquad (23)$$

The *dissociative mechanism* involves the rate-determining (slow) loss of Y to give a five-coordinated intermediate. The subsequent addition of Z to the intermediate is rapid (24). This is called an S_N1 process, meaning *substitution, nucleophilic, unimolecular*. The reaction is nucleophilic because the incoming ligand seeks a positive center (like the nucleus of an atom), the metal ion. *Unimolecular reactions* are those in which the rate-determining step involves only one reactant species.

The *displacement or* S_N2 *mechanism* involves the formation of a seven-coordinated intermediate in a slow step and its subsequent dissociation in a fast step (25). The reaction is *bimolecular;* two reactants are involved in the slow step. These two mechanisms can

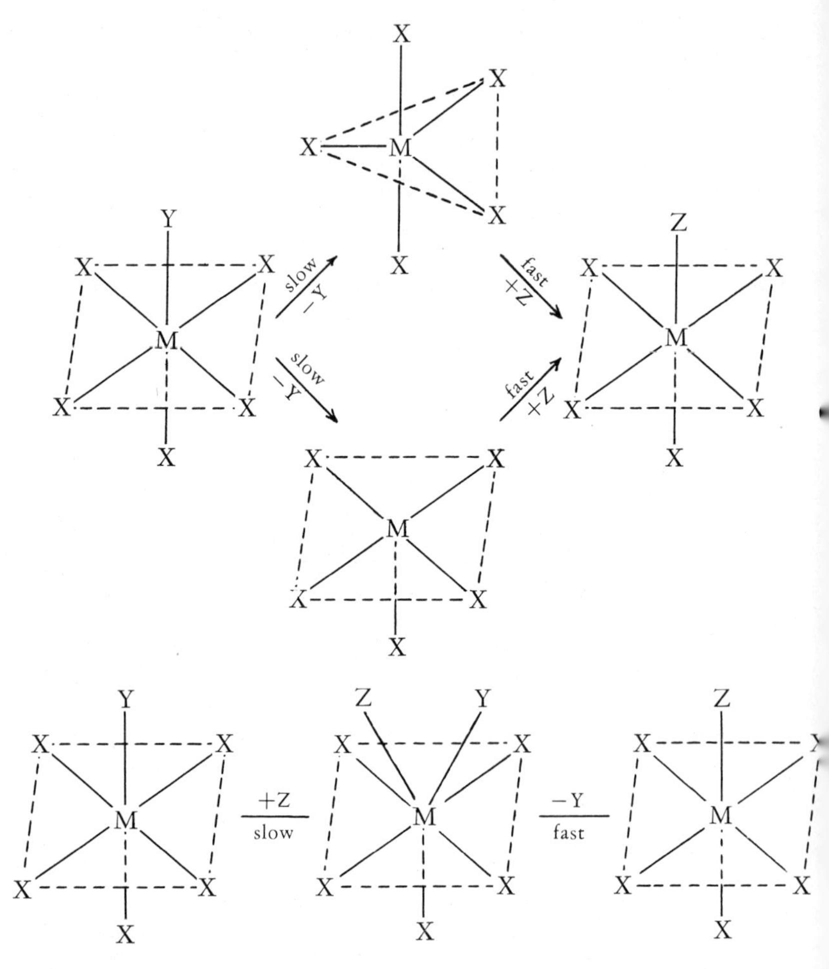

be differentiated by noting that in an S_N1 process the important feature is bond-breaking; in an S_N2 reaction the formation of an additional metal ligand bond is of an importance equal to or greater than that of the breaking of a metal-ligand bond. The S_N1 or S_N2 designation therefore indicates the relative importance of bond making and breaking in the rate-determining step of a reaction. Let us now look at several systems that have been studied to see how mechanistic information on substitution reactions of coordination compounds was obtained.

6–6 OCTAHEDRAL SUBSTITUTION REACTIONS

The most fundamental substitution reaction in aqueous solution, water exchange (22), has been studied for a variety of metal ions (Figure 6–3). The exchange of water in the coordination sphere of a metal with bulk solvent water occurs very rapidly for most metal ions, and therefore the rates of these reactions were studied primarily by relaxation techniques. In these methods a system at equilibrium is disturbed, for example, by a very sudden increase in temperature. Under the new condition—the higher temperature—the system will no longer be at equilibrium. The rate of equilibration can then be measured. If one can change the temperature of a solution in 10^{-8} sec, then one can measure the rates of reactions that take longer than 10^{-8} sec.

The rate of replacement of coordinated water molecules by $SO_4^=$, $S_2O_3^=$, EDTA, and other species has also been measured for a variety of metal ions (26). The rates of these reactions are dependent on the

$$[M(H_2O)_x]^{n+} + L^{2-} \longrightarrow [M(H_2O)_{x-1}L]^{(n-2)+} + H_2O \qquad (26)$$

concentration of hydrated metal ion, but they are independent of the concentration of the entering ligand, that is, a first-order rate law, Equation (27), applies. In many cases the rate of reaction (26) for a

$$\text{rate} = k[M(H_2O)_x{}^{n+}] \qquad (27)$$

given metal ion is independent of whether H_2O, $SO_4^=$, $S_2O_3^=$, or EDTA is the entering ligand (L). This observation and the fact that the rate law does not include the entering ligand suggest that these reactions occur by a mechanism in which the slow step is the breaking of a bond between the metal ion and water. The resulting species would then be expected to coordinate rapidly with any nearby species.

As was suggested in Section 6–4, the fact that the more highly charged hydrated metal ions such as Al^{3+} and Sc^{3+} undergo H_2O exchange more slowly than M^{2+} and M^+ ions also suggests that bond breaking is important in the rate-determining step of these reactions.

The evidence gathered in these studies is not conclusive, but it does suggest that S_N1 processes are important in substitution reactions of hydrated metal ions.

Probably the most widely studied coordination compounds are the ammine complexes of cobalt(III). Their stability, ease of preparation, and slow reactions makes them particularly amenable to kinetic study. Since work on these complexes has been done almost exclusively in water, the reactions of the complexes with the solvent water had to be considered first. In general, ammonia or amines coordinated to Co(III) are observed to be replaced so slowly by water that only the replacement of ligands other than amines is usually considered.

The rates of reactions of the type (28) have been studied and

$$[Co(NH_3)_5X]^{2+} + H_2O \rightarrow [Co(NH_3)_5OH_2]^{3+} + X^- \quad (28)$$

found to be first order in the cobalt complex (X can be any of a variety of anions). Since in aqueous solution the concentration of H_2O is always about 55.5 M, the effect of changes in water concentration on the reaction rate cannot be determined. Rate laws (29) and (30)

$$\text{rate} = k[Co(NH_3)_5X^{2+}] \quad (29)$$

$$\text{rate} = k'[Co(NH_3)_5X^{2+}][H_2O] \quad (30)$$

are experimentally indistinguishable in aqueous solution, since k may simply be $k'[H_2O] = k'[55.5]$. Therefore, the rate law does not tell us whether H_2O is involved in the rate-determining step of the reaction. The decision as to whether these reactions proceed by an S_N2 displacement of X by H_2O or by an S_N1 dissociation followed by addition of H_2O must be made from other experimental data.

Two types of experiments have provided good mechanistic evidence. The rate of hydrolysis (replacement of one chloride by water) of *trans*-$[Co(NH_3)_4Cl_2]^+$ is approximately 10^3 times faster than that of $[Co(NH_3)_5Cl]^{2+}$. Increased charge on a complex is expected to strengthen metal-ligand bonds and hence retard metal-ligand bond cleavage. It is also expected to attract incoming ligands and aid displacement reactions. Since a decrease in rate is observed as the charge on the complex increases, a dissociative (S_N1) process seems to be operative.

Another piece of evidence resulted from the study of the hydrolyses of a series of complexes related to *trans*-$[Co(en)_2Cl_2]^+$. In these complexes the ethylenediamine was replaced by similar diamines in which H atoms on C were replaced by CH_3 groups. The complexes containing the substituted diamines react more rapidly than the ethylenediamine complex. The replacement of H by CH_3 increases the bulk of the ligands. From models of these compounds one can see that this should make it more difficult for an attacking ligand to approach the metal atom. This steric crowding should retard an S_N2 reaction. By crowding the vicinity of the metal atom with bulky ligands, one enhances a dissociative process, since the removal of one ligand reduces the congestion around the metal. The increase in rate observed when the more bulky ligands were used is good evidence for the S_N1 mechanism.

As a result of a large number of studies on acido amine complexes of cobalt(III), it appears that replacement of the acido group by H_2O occurs by a process that is primarily dissociative in character. The ligand-cobalt bond is stretched to some critical distance before a H_2O molecule begins to enter the coordination sphere. In complexes that have a charge of 2+ or greater the breaking of a cobalt-ligand bond becomes quite difficult, and the entering H_2O molecule plays a more important role.

Replacement of an acido group (X^-) in a cobalt(III) complex with a group other than H_2O (31) has been observed to take place by

$$[Co(NH_3)_5X]^{2+} + Y^- \longrightarrow [Co(NH_3)_5Y]^{2+} + X^- \qquad (31)$$

initial substitution by solvent H_2O with subsequent replacement of water by the new group Y (32). Therefore, in a number of co-

$$[Co(NH_3)_5X]^{2+} \xrightarrow[\text{slow}]{H_2O} [Co(NH_3)_5H_2O]^{3+} \xrightarrow[\text{fast}]{Y} [Co(NH_3)_5Y]^{2+} \qquad (32)$$

balt(III) reactions the rates of reaction (31) are the same as the rate of hydrolysis (28).

Hydroxide ion is uniquely different from other reagents with respect to its reactivity toward Co(III) ammine complexes. It reacts very rapidly (as much as 10^6 times faster than H_2O) with cobalt(III) ammine complexes in a *base hydrolysis* reaction (33). In this reaction

$$[Co(NH_3)_5Cl]^{2+} + OH^- \rightarrow [Co(NH_3)_5OH]^{2+} + Cl^- \quad (33)$$

a first-order dependence on the substituting ligand OH^- is observed (34). The second-order kinetics and the unusually rapid reaction

$$\text{rate} = k[Co(NH_3)_5Cl^{2+}][OH^-] \quad (34)$$

suggest that OH^- is an exceptionally good nucleophilic reagent toward Co(III) and that the reaction proceeds through an S_N2-type intermediate. However, an alternative mechanism, (35), (36), (37),

$$[Co(NH_3)_5Cl]^{2+} + OH^- \overset{\text{fast}}{\rightleftharpoons} [Co(NH_3)_4NH_2Cl]^+ + H_2O \quad (35)$$

$$[Co(NH_3)_4NH_2Cl]^+ \xrightarrow{\text{slow}} [Co(NH_3)_4NH_2]^{2+} + Cl^- \quad (36)$$

$$[Co(NH_3)_4NH_2]^{2+} + H_2O \xrightarrow{\text{fast}} [Co(NH_3)_5OH]^{2+} \quad (37)$$

will also explain this behavior. In reaction (35), $[Co(NH_3)_5Cl]^{2+}$ acts as a Brønsted acid to give $[Co(NH_3)_4NH_2Cl]^+$, which is known as an *amido* ($:\ddot{N}H_2^-$ containing) compound and which is the conjugate base of $[Co(NH_3)_5Cl]^{2+}$. The reaction then proceeds by an S_N1 process (36) to give a five-coordinated intermediate which then reacts with the abundant solvent molecules to give the observed product (37). This mechanism is consistent with second-order rate behavior, and yet it involves an S_N1 mechanism. Since the reaction involves the conjugate base of the initial complex in an S_N1 rate-determining step, the symbol S_N1CB has been given to the mechanism.

The task of determining which of these mechanisms best explains the experimental observations is very difficult. However, convincing evidence that has been presented supports the S_N1CB hypothesis. Among the best arguments in favor of the mechanism are the following: Octahedral Co(III) complexes in general react primarily by a dissociation process, and there is no convincing reason why OH^- should initiate an S_N2 process. Hydroxide ion has been found to be a poor nucleophile in reactions with Pt(II), and therefore it seems unreasonable that it should be unusually reactive toward Co(III). In reactions of Co(III) in nonaqueous solutions there is excellent evidence for the presence of the five-coordinated intermediate proposed by the S_N1CB mechanism.

A final important piece of evidence is that if no N—H hydrogen is present in a Co(III) complex, the complex reacts slowly with OH^-. This certainly suggests that the acid-base properties of the complex are more important to the rate of reaction than the nucleophilic properties of OH^-. This base hydrolysis reaction of Co(III) ammine complexes illustrates the fact that kinetic data often can be interpreted in more than one way and that rather subtle experiments must be performed to eliminate one or more possible mechanisms.

Substitution reactions of a wide variety of octahedral compounds have now been studied. Where mechanistic interpretation of the data has been made, a dissociative-type process has most frequently been postulated. This result should not be surprising, since six ligands around a central atom leave little room for adding another group. In a very few examples, evidence for seven-coordinated intermediates or for influence of the entering ligand has been presented. Therefore, the S_N2 mechanism cannot be discarded as a conceivable path for octahedral substitution.

6–7 SQUARE PLANAR SUBSTITUTION

Complexes in which the coordination number of the metal is less than six would be more likely to react by a displacement S_N2 process. Of the complexes in which the coordination number of the metal is less than six, the four-coordinated Pt(II) complexes have been most thoroughly studied. Experimental evidence for an S_N2 mechanism has been presented. The rates of reaction for some Pt(II) complexes having different charges are shown in Table 6–2. In the series of complexes in Table 6–2 the charge on the reactant Pt(II) complex goes from -2 to $+1$, and yet the rate changes by a factor of only two (quite a small effect). The breaking of a Pt—Cl bond should become

TABLE 6-2

The Rates of Some Reactions of Pt(II) Complexes

Reaction	$t_{\frac{1}{2}}$ *at 25° C, min*
$[PtCl_4]^{2-} + H_2O \rightarrow [PtCl_3H_2O]^- + Cl^-$	300
$[PtNH_3Cl_3]^- + H_2O \rightarrow [PtNH_3Cl_2H_2O]^0 + Cl^-$	310
cis-$[Pt(NH_3)_2Cl_2]^0 + H_2O \rightarrow [Pt(NH_3)_2ClH_2O]^+ + Cl^-$	300
$[Pt(NH_3)_3Cl]^+ + H_2O \rightarrow [Pt(NH_3)_3H_2O]^{2+} + Cl^-$	690

much more difficult as the charge on the complex becomes more positive; however, the formation of a new bond should become more favorable. The small effect of charge on the complex on reaction rate suggests that both bond making and bond breaking are important, as is characteristic in an S_N2 process.

Good evidence for the importance of the entering ligand would be second-order kinetics, first order in Pt(II) complex and first order in entering ligand. This in fact was found for the reactions of many different platinum(II) complexes with a variety of different ligands. There is a slight complication because the solvent water also behaves as a potential ligand. The result is that reactions such as (38) obey

$$[Pt(NH_3)_3Cl]^+ + Br^- \xrightarrow{H_2O} [Pt(NH_3)_3Br]^+ + Cl^- \qquad (38)$$

a two-term rate law (39). This type of rate law indicates the re-

$$\text{rate} = k[Pt(NH_3)_3Cl^+] + k'\,[Pt(NH_3)_3Cl^+][Br^-] \qquad (39)$$

action is occurring by two mechanistic paths, only one of which involves Br^- in the rate-determining step.

Since experience with platinum(II) reactions suggests that the Br^- independent path is not an S_N1 process, it has been postulated that the solvent H_2O replaces Cl^- in a slow step and is subsequently replaced in a rapid step by Br^-. The postulated mechanism is presented in Figure 6–4. That the solvent can participate in this way was demonstrated in experiments in which a similar reaction was carried out in variety of different solvents. In solvents that are poor ligands (CCl_4, C_6H_6) second-order kinetic behavior is observed and the entering ligand presumably enters the complex directly; in solvents that are good ligands (H_2O, alcohols) the first-order path also contributes to the reaction.

In the substitution reactions of both square planar platinum(II) complexes and octahedral cobalt(III) complexes the influence of the solvent is quite marked. It should be realized that in all solution processes the solvent plays an important role. Thus the behavior observed in H_2O may differ markedly from that found in other solvents.

That square planar Pt(II) complexes react by an S_N2 mechanism is now widely accepted. Although little kinetic work has been done on substitution reactions of other square complexes, it seems probable that S_N2 processes will also predominate in these systems. When the entering ligand plays a role in determining the rate of a reaction, it is important to learn which ligands promote the most rapid reactions.

Rate studies have indicated that ligands that exhibit a large trans effect (Section 4–8) also add rapidly to platinum(II) complexes. Groups such as phosphines, SCN^-, and I^- react rapidly with

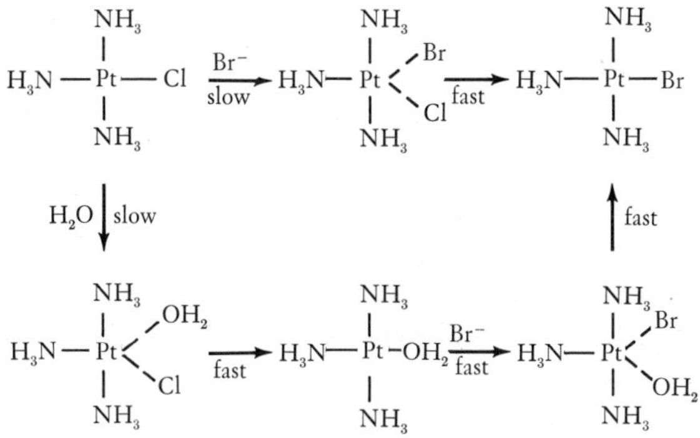

$$[Pt(NH_3)_3Cl]^{+} \xrightarrow[\text{slow}]{Br^-} [Pt(NH_3)_3(Br)(Cl)] \xrightarrow{\text{fast}} [Pt(NH_3)_3Br]^{+}$$

$$[Pt(NH_3)_3Cl]^{+} \xrightarrow[\text{slow}]{H_2O} [Pt(NH_3)_3(OH_2)(Cl)] \xrightarrow{\text{fast}} [Pt(NH_3)_3(OH_2)]^{2+} \xrightarrow[\text{fast}]{Br^-} [Pt(NH_3)_3(Br)(OH_2)] \xrightarrow{\text{fast}} [Pt(NH_3)_3Br]^{+}$$

Figure 6–4 The mechanism for reaction (38).

Pt(II) complexes; amines, Br^-, and Cl^- react at an intermediate rate; and H_2O and OH^- react slowly. This behavior reflects in part the *nucleophilicity* (the attraction toward a positive center) of these groups and indicates that OH^- is a poor nucleophile, at least toward Pt(II). However, the order of reactivity is clearly not an indication of only the attraction of ligands toward a positive center. If it were, Cl^- should certainly react more rapidly than the larger anions Br^- and I^-. The observed order of reactivity can be related to the ease with which the entering ligand can release its electrons to Pt(II). Iodide ion is a better releasing agent than is Cl^-. A reasonably good correlation between the reactivity of the entering group and its oxidation potential is found. In general the more easily a group is oxidized the more rapidly it reacts with platinum(II) complexes.

Only a limited number of rate and mechanistic studies of reactions of tetrahedral metal complexes have been made. These complexes are uncommon compared to octahedral species, and their substitution reactions are frequently very rapid. It is quite possible that the rapid reactions are at least a partial indication that S_N2 processes are occurring. The energy requirement (activation energy) for a reaction will be reduced when an incoming group can assist in the cleavage of a metal-ligand bond.

6–8 MECHANISMS FOR REDOX REACTIONS

Let us now consider redox reactions, the other category of reactions of coordination compounds. Redox reactions are those in which the oxidation states of some atoms change. In reaction (40)

$$[Co(NH_3)_5Cl]^{2+} + [Cr(H_2O)_6]^{2+} + 5H_3O^+ \rightarrow [Co(H_2O)_6]^{2+} + [Cr(H_2O)_5Cl]^{2+} + 5NH_4^+ \quad (40)$$

the oxidation state of Co changes from 3+ to 2+ (Co is reduced); the oxidation state of chromium increases from 2+ to 3+ (Cr is oxidized). This change of oxidation state implies that an electron was transferred from Cr(II) to Co(III) (41), (42). The mechanism

$$Cr(II) \rightarrow Cr(III) + e \quad (41)$$

$$e + Co(III) \rightarrow Co(II) \quad (42)$$

for this reaction should suggest how the electron is transferred. Two basic paths appear to be possible. In one the electron effectively hops from one species to the other. This is called the *electron transfer* or *outer-sphere activated complex* mechanism. In the other process the oxidant and reductant are attached to each other by a bridging molecule, atom, or ion through which the electron can pass. This is called the *atom transfer* or *bridged activated complex* mechanism.

Elegant experiments that demonstrated the validity of the atom transfer path were performed by Taube and his coworkers. Reaction (40) was one of many studied. It was observed that in the reduction of $[Co(NH_3)_5Cl]^{2+}$ by Cr^{2+} the chromium(III) product always contained a chloride ion. In more detailed studies $[Co(NH_3)_5Cl]^{2+}$ containing radioactive $^{36}Cl^-$ was dissolved in a solution containing Cr^{2+} and unlabeled Cl^-. After the reduction, which is very rapid, the product $[Cr(H_2O)_5Cl]^{2+}$ was examined and found to contain only labeled $^{36}Cl^-$. This proved that the cobalt complex was the only source of the Cl^- that was eventually found in the chromium(III) complex. To explain these results, a mechanism in which the activated complex contains cobalt and chromium atoms linked by a chloride ion was proposed (I). The chloride bridge provides a good

$$\left[(H_3N)_5Co\text{—}Cl\text{—}Cr(OH_2)_5\right]^{4+}$$

I

path between the two metal atoms for electron transfer, much as a copper wire connecting two electrodes provides a path. Once an electron is transferred from Cr(II) to Co(III), the Cr(III) formed attracts the Cl^- more strongly than does Co(II), and therefore the Cl^- becomes part of the Cr(III) complex. A direct electron transfer from the chromium complex to the cobalt complex followed by the transfer of the $^{36}Cl^-$ seems unlikely. If this were the path, then unlabeled Cl^- from the solution would be as readily incorporated into the Cr(III) complex as is the $^{36}Cl^-$ attached to cobalt.

Reaction (40) and similar reactions were very clever choices for this study because Co(III) and Cr(III) complexes are inert, whereas Cr(II) and Co(II) complexes are labile. Thus the rapid redox reaction is complete long before any substitution reactions begin to occur on the Co(III) and Cr(III) complexes. The lability of $[Cr(H_2O)_6]^{2+}$ allows the complex to lose a molecule of water rapidly and form the activated bridged intermediate, I. The results obtained demand a mechanism in which the coordinated chloride ion never escapes alone into the solution; for in that case appreciable quantities of $[Cr(H_2O)_6]^{3+}$ and unlabeled $[Cr(H_2O)_5Cl]^{2+}$ would be formed. A mechanism in which $^{36}Cl^-$ is attached to both Cr and Co during the electron transfer appears to fit the experimental data very well.

Reductions of a series of cobalt(III) complexes, $[Co(NH_3)_5X]^{2+}$, with chromous solutions have been studied. The transfer of the X^- group to chromium occurred when X^- was NCS^-, N_3^-, PO_4^{3-}, $C_2H_3O_2^-$, Cl^-, Br^-, and $SO_4^=$ (43). This suggests that all of these

$$[Co(NH_3)_5X]^{2+} + [Cr(H_2O)_6]^{2+} + 5H_3O^+ \rightarrow [Co(H_2O)_6]^{2+} + [Cr(H_2O)_5X]^{2+} + 5NH_4^+ \quad (43)$$

reactions occur by the atom transfer mechanism. The rates of these reactions increased in the order $C_2H_3O_2^- < SO_4^= < Cl^- < Br^-$. Presumably those ions that form bridges most readily and those that provide the best path for electrons produce the fastest reactions. It is of interest to note that the complex

$$[(NH_3)_5Co{-}O{-}\overset{\overset{\displaystyle O}{\|}}{C}{-}CH{=}CH\overset{\overset{\displaystyle O}{\|}}{C}{-}OH]^{2+}$$

is readily reduced by Cr(II), whereas the reduction of

$$[(NH_3)_5Co{-}O\overset{\overset{\displaystyle O}{\|}}{C}CH_2CH_2\overset{\overset{\displaystyle O}{\|}}{C}{-}OH]^{2+}$$

is much slower. The difference is believed to be due to the fact that although both groups should form bridges between Co and Cr, the organic molecule that contains the carbon-carbon double bond is a much better conductor of electrons.

Redox reactions that proceed by electron transfer through a bridging group are quite common. In the reactions cited, transfer of the bridging atom followed the redox reaction. This is not a necessary result of the mechanism, but in its absence it is difficult to determine whether or not a bridging atom was involved in the electron transfer process. There are a variety of redox reactions of metal complexes that probably occur by direct electron transfer. The rate of the redox reaction (44) (which is actually no reaction at all) can be studied by labeling either of the complexes with a radioactive isotope

$$[{}^{*}Fe(CN)_6]^{4-} + [Fe(CN)_6]^{3-} \rightarrow [{}^{*}Fe(CN)_6]^{3-} + [Fe(CN)_6]^{4-} \quad (44)$$

of Fe or with ^{14}C; the reaction is very rapid.

Both ferrocyanide and ferricyanide ions are inert ($[Fe(CN)_6]^{4-}$ is a low-spin d^6 system; $[Fe(CN)_6]^{3-}$ is a low-spin d^5 system); therefore, loss or exchange of CN^- or any substitution reaction is very slow. The fact that the redox reaction is very fast whereas substitution reactions are very slow essentially eliminates the possibility of electron transfer through a bridged activated complex, since formation of the activated complex amounts to a substitution process.

When one rules out the bridge mechanism, one is left with direct electron transfer. On theoretical grounds there is a critical requirement for such a process. The *Franck-Condon principle* states that there can be no appreciable change of atomic arrangement during the time of an electronic transition, that is, very light electrons move much more rapidly than the much heavier atoms. Let us consider the effect of this on a direct electron transfer process. The ligands can more closely approach the smaller Fe^{3+} ion than the larger Fe^{2+}(II). During the transfer of an electron from $[Fe(CN)_6]^{4-}$ to $[Fe(CN)_6]^{3-}$, none of the Fe, C, or N atoms move. The result of electron transfer is therefore the formation of an $[Fe(CN)_6]^{3-}$ in which the Fe—C bonds are too long and an $[Fe(CN)_6]^{4-}$ in which the Fe—C bonds are too short. Both of the products are of a higher energy than the normal

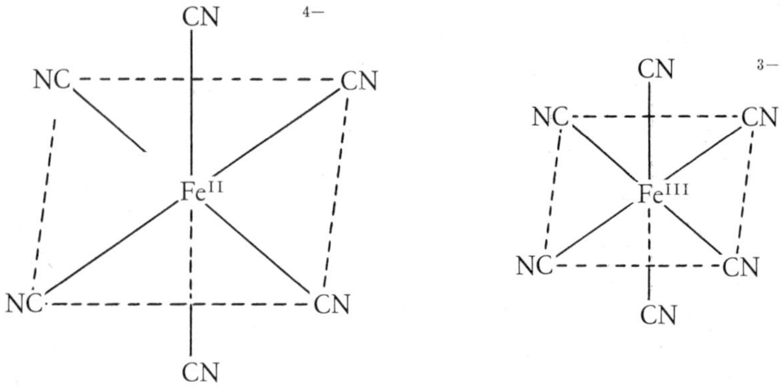

II

ions in which the Fe—C bonds have their proper length (the length that gives the lowest energy).

The process described here is an example of a perpetual-motion machine. We took $[Fe(CN)_6]^{3-}$ and $[Fe(CN)_6]^{4-}$ ions, transferred an electron, and immediately obtained the same two species but with each now having an excess of energy. A process in which there is a net gain in energy cannot take place; hence this description of the reaction must be wrong. The reaction can occur only if we impart at least as much energy as we remove. Therefore, before electron transfer will occur, the Fe—C bonds in $[Fe(CN)_6]^{4-}$ must become shorter, the Fe—C bonds in $[Fe(CN)_6]^{3-}$ must become longer; and for this to be so, energy must be added to the system. For this reaction a suitable configuration for reaction would be one in which the $[Fe(CN)_6]^{3-}$ and $[Fe(CN)_6]^{4-}$ ions have equivalent geometries. Then the products and reactants in the electron transfer process would be equivalent, and no energy would be produced as a result of the electron transfer.

One can understand the rates of many direct electron transfer reactions by considering the amount of energy necessary to make the oxidant and reductant look alike. Since $[Fe(CN)_6]^{3-}$ and $[Fe(CN)_6]^{4-}$ are rather similar, a relatively small addition of energy (the activation energy) will make the ions alike; thus electron transfer can occur rapidly. Reaction (45) is very slow. The complexes $[Co(NH_3)_6]^{2+}$ and $[Co(NH_3)_6]^{3+}$ do not differ greatly in size, and

$$[^*Co(NH_3)_6]^{3+} + [Co(NH_3)_6]^{2+} \rightarrow [^*Co(NH_3)_6]^{2+} + [Co(NH_3)_6]^{3+} \quad (45)$$

hence one might expect electron exchange between these two complexes to be rapid. The two complexes do, however, differ in electronic configuration; the $[Co(NH_3)_6]^{2+}$ is $t_{2g}{}^5e_g{}^2$, $[Co(NH_3)_6]^{3+}$ is $t_{2g}{}^6$. Hence, both the length of the Co—N bonds and the electronic configurations must change prior to electron transfer. This is the reason for the very slow reaction.

Other factors also influence the rate of direct electron transfer processes. For example, the greater the conductivity of their ligands the more readily should electron transfer proceed between two complexes. Cyanide ions would be expected to provide a good path for electrons, and indeed electron transfer between a variety of similar cyanide complexes has been found to be rapid. The same is true of the highly conducting systems $[M(phen)_3]^{n+}$ and $[M(bipy)_3]^{n+}$ relative to $[M(en)_3]^{n+}$ and $[M(NH_3)_6]^{n+}$.

PROBLEMS

1. Designate whether the following complexes are expected to be inert or labile and give a reason for your choice.

$[Al(C_2O_4)_3]^{3-}$
$[Cr(C_2O_4)_3]^{3-}$
$[CoF_6]^{3-}$ (high-spin)
$[Fe(CN)_6]^{4-}$ (low-spin)
$[V(H_2O)_6]^{3+}$
$[V(H_2O)_6]^{2+}$
$[Ni(NH_3)_6]^{2+}$
$[PtCl_6]^{2-}$ (low-spin)

2. For analogous complexes of each of the following series of metal ions, indicate the order of decreasing lability and explain your answer. (*a*) Mg^{2+}, Ca^{2+}, Sr^{2+}, Ba^{2+}, and Ra^{2+}. (*b*) Mg^{2+}, Al^{3+}, and Si^{4+}. (*c*) High-spin Ca^{2+}, V^{2+}, Cr^{2+}, Mn^{2+}, Fe^{2+}, Co^{2+}, Ni^{2+}, Cu^{2+}, and Zn^{2+}.

3. Explain why $[Co(NH_3)_6]^{3+}$ is reduced much more slowly than $[Co(NH_3)_5Cl]^{2+}$ by $[Cr(H_2O)_6]^{2+}$. Write the formula of the Cr(III) reaction product in each case.

4. The complex $[Co(NH_3)_5SO_3]^+$ will react in acid solution to generate SO_2. One might expect the rate law for this reaction to have the form

$$\text{rate} = k[\text{Co(NH}_3)_5\text{SO}_3^+][\text{H}^+]$$

$$[\text{Co(NH}_3)_5\text{OSO}_2]^+ + 2\text{H}^+ \xrightarrow{\text{H}_2{}^{18}\text{O}} [\text{Co(NH}_3)_5\text{OH}_2]^{3+} + \text{SO}_2$$

It is probable that the reaction will occur without the incorporation of ^{18}O into the products from labeled solvent $H_2{}^{18}O$. Propose a reasonable mechanism that is consistent with this rate law and the absence of labeled ^{18}O in the products.

5. Both of the complexes $[AuCl_4]^-$ and $[Au(dien)Cl]^{2+}$ undergo very rapid radiochloride exchange:

$$[\text{Au(dien)Cl}]^{2+} + {}^*\text{Cl}^- \rightarrow [\text{Au(dien)}^*\text{Cl}]^{2+} + \text{Cl}^-$$

The rate laws for these reactions have the form

$$\text{rate} = k[\text{complex}] + k'[\text{complex}][\text{Cl}^-]$$

Propose a reasonable mechanism that is consistent with these data.

6. The $[M(bipy)_3]^{2+}$ complexes of the first-row transition metals dissociate in water. The rates of these reactions have been measured

$$[\text{M(bipy)}_3]^{2+} + 2\text{H}_2\text{O} \rightarrow [\text{M(bipy)}_2(\text{H}_2\text{O})_2]^{2+} + \text{bipy}$$

and found to increase in the order $Fe^{2+} < Ni^{2+} < Co^{2+} \sim Mn^{2+} \sim Cu^{2+} \sim Zn^{2+}$. Propose a reason for this order of reactivity.

REFERENCES

F. Basolo and R. G. Pearson, *Mechanisms of Inorganic Reactions*, Wiley-Interscience, New York, 1958.

H. Taube, "Rates and mechanisms of substitution in inorganic complexes in solution," *Chem. Revs.*, **50,** 69 (1952).

H. Taube, "Mechanisms of redox reactions of simple chemistry," in H. J. Emeléus and A. G. Sharpe (eds.), *Advances in Inorganic Chemistry and Radiochemistry*, Academic, New York, 1959, vol. I, pp. 1–50.

C. K. Ingold, R. S. Nyholm, and M. L. Tobe, "Orienting effects in octahedral aquation," *Nature*, **187,** 477 (1960); "Orientation in octahedral basic hydrolysis," *Nature*, **194,** 344 (1962).

R. G. Pearson, "Crystal field theory and substitution reactions of metal ions," *J. Chem. Educ.*, **38,** 164 (1961).

F. Basolo and R. G. Pearson, "Mechanisms of substitution reactions of metal complexes," in H. J. Emeléus and A. G. Sharpe (eds.), *Advances in Inorganic Chemistry and Radiochemistry*, Academic, New York, 1961, vol. III, pp. 1–89; "The *trans* effect in metal complexes," in F. A. Cotton (ed.), *Progress in Inorganic Chemistry*, Wiley-Interscience, New York, vol. IV, pp. 381–453.

N. Sutin, "Electron exchange reactions," *Ann. Rev. Nuclear Sci.*, **12,** 285 (1962).

Index of Complexes

Index

813 6